PRIVACY AND IDENTITY PROTECTION

# AUTOMATED FINGERPRINT IDENTIFICATION

# BACKGROUND AND INTEROPERABILITY STRATEGY

# PRIVACY AND IDENTITY PROTECTION

Additional books in this series can be found on Nova's website
under the Series tab.

Additional e-books in this series can be found on Nova's website
under the e-book tab.

PRIVACY AND IDENTITY PROTECTION

# AUTOMATED FINGERPRINT IDENTIFICATION

## BACKGROUND AND INTEROPERABILITY STRATEGY

EDWARD P. JACKSON
EDITOR

nova publishers
*New York*

## NOTICE TO THE READER

The Publisher has taken reasonable care in the preparation of this book, but makes no expressed or implied warranty of any kind and assumes no responsibility for any errors or omissions. No liability is assumed for incidental or consequential damages in connection with or arising out of information contained in this book. The Publisher shall not be liable for any special, consequential, or exemplary damages resulting, in whole or in part, from the readers' use of, or reliance upon, this material. Any parts of this book based on government reports are so indicated and copyright is claimed for those parts to the extent applicable to compilations of such works.

Independent verification should be sought for any data, advice or recommendations contained in this book. In addition, no responsibility is assumed by the publisher for any injury and/or damage to persons or property arising from any methods, products, instructions, ideas or otherwise contained in this publication.

This publication is designed to provide accurate and authoritative information with regard to the subject matter covered herein. It is sold with the clear understanding that the Publisher is not engaged in rendering legal or any other professional services. If legal or any other expert assistance is required, the services of a competent person should be sought. FROM A DECLARATION OF PARTICIPANTS JOINTLY ADOPTED BY A COMMITTEE OF THE AMERICAN BAR ASSOCIATION AND A COMMITTEE OF PUBLISHERS.

Additional color graphics may be available in the e-book version of this book.

**Library of Congress Cataloging-in-Publication Data**

ISBN: 978-1-63483-304-2

*Published by Nova Science Publishers, Inc. † New York*

# CONTENTS

# PREFACE

From its inception more than 100 years ago, fingerprint technology has come a long way. This book provides background on and traces the recent history of automated fingerprint identification systems. Included in the book is a brief examination of standards and the role of algorithms. In its history, fingerprints have been used to identify criminal and terrorist suspects, perform background checks, and monitor immigration status, among other applications. Prints found at the scene of a crime, referred to as latent prints, are compared with fingerprint records of known individuals who have been convicted of or arrested for a crime. Law enforcement agencies have been working to automate this comparison process. This book also provides the current landscape of latent Automated Fingerprint Identification System (AFIS) interoperability, meaning the ability to share information between networks, systems, components, etc. The book addresses essential elements of interoperability, outlines recent advances in the field and provides a high-level strategic plan for Federal agencies to implement the necessary requirements for interoperability to enhance public safety.

In: Automated Fingerprint Identification
Editor: Edward P. Jackson
ISBN: 978-1-63483-304-2


*Chapter 1*

# AUTOMATED FINGERPRINT IDENTIFICATION SYSTEM (AFIS)*

***Kenneth R. Moses, Peter Higgins, Michael Mccabe, Salil Prabhakar and Scott Swann***

## 1. INTRODUCTION

Prior to the industrial revolution and the mass migrations to the cities, populations lived mostly in rural communities where everyone knew everyone else and there was little need for identification. Indeed, there were no police forces, no penitentiaries, and very few courts. As cities became crowded, crime rates soared and criminals flourished within a sea of anonymity. Newspapers feasted on stories of lawlessness, legislatures quickly responded with more laws and harsher penalties (especially for repeat offenders), and police departments were charged with identifying and arresting the miscreants. Identification systems—rogues' galleries, anthropometry, Bertillon's "portrait parlé", and the Henry system—emerged and quickly spread worldwide at the end of the 19th and beginning of the 20th century.

The late 1960s and early 1970s witnessed another era of civil turmoil and an unprecedented rise in crime rates, but this era happened to coincide with the

* This is an edited, reformatted and augmented version of a document that originally appeared as Chapter 6 in The Fingerprint Sourcebook, issued by the National Institute of Justice, July 2011.

development of the silicon chip. The challenges inherent in identification systems seemed ready-made for the solutions of automatic data processing, and AFIS—Automated Fingerprint Identification System—was born.

During this same period, The RAND Corporation, working under a national grant, published *The Criminal Investigative Process* (Greenwood et al., 1975), a comprehensive study and critique of the process by which crimes get solved—or do not. Generally critical of traditional methods used by detectives, the study placed any hopes for improvement on physical evidence in general and latent prints in particular. In a companion study, Joan Petersilia concluded that:

> No matter how competent the evidence technician is at performing his job, the gathering of physical evidence at a crime scene will be futile unless such evidence can be properly processed and analyzed. Since fingerprints are by far the most frequently retrieved physical evidence, making the system of analyzing such prints effective will contribute the most toward greater success in identifying criminal offenders through the use of physical evidence. (Petersilia, 1975, p 12)

Though new technology was already in development at the Federal Bureau of Investigation (FBI), it would be a popular movement at the local and state levels that would truly test Petersilia's theory.

## 1.1. Need for Automation

In 1924, the FBI's Identification Division was established by authority of the United States congressional budget appropriation bill for the Department of Justice. The identification division was created to provide a central repository of criminal identification data for law enforcement agencies throughout the United States. The original collection of fingerprint records contained 810,188 records. After its creation, hundreds of thousands of new records were added to this collection yearly, and by the early 1960s the FBI's criminal file had grown to about 15 million individuals. This was in addition to the 63 million records in the civilian file, much of which was the result of military additions from World War II and the Korean conflict.

Almost all of the criminal file's 15 million individuals contained 10 rolled fingerprints per card for a total of 150 million single fingerprints. Incoming records were manually classified and searched against this file using the FBI's modified Henry system of classification. Approximately 30,000 cards were searched daily. The time and human resources to accomplish this daily workload continued to grow. As a card entered the system, a preliminary gross pattern classification was assigned to each fingerprint by technicians. The technicians could complete approximately 100 fingerprint cards per hour. Complete classification and searching against the massive fi les could only be accomplished at an average rate of 3.3 cards per employee per hour. Obviously, as the size of the criminal file and the daily workload increased, the amount of resources required continued to grow. Eventually, classification extensions were added to reduce the portion of the criminal file that needed to be searched against each card. Nonetheless, the manual system used for searching and matching fingerprints was approaching the point of being unable to handle the daily workload.

Although punch card sorters could reduce the number of fingerprint cards required to be examined based on pattern classification and other parameters, it was still necessary for human examiners to scrutinize each fingerprint card on the candidate list. A new paradigm was necessary to stop the increasing amount of human resources required to process search requests. A new automated approach was needed to (1) extract each fingerprint image from a tenprint card, (2) process each of these images to produce a reduced-size template of characteristic information, and (3) search a database to automatically produce a highly reduced list of probable candidate matches (Cole, 2001, pp 251–252).

## 1.2. Early AFIS Development

In the early 1960s, the FBI in the United States, the Home Office in the United Kingdom, Paris Police in France, and the Japanese National Police initiated projects to develop automated fingerprint identification systems. The thrust of this research was to use emerging electronic digital computers to assist or replace the labor-intensive processes of classifying, searching, and matching tenprint cards used for personal identification.

## 1.3. FBI AFIS Initiative

By 1963, Special Agent Carl Voelker of the FBI's Identification Division realized that the manual searching of the criminal file would not remain feasible for much longer. In an attempt to resolve this problem, he sought the help of engineers Raymond Moore and Joe Wegstein of the National Institute of Standards and Technology (NIST).[1] After describing his problem, he asked for assistance in automating the FBI's fingerprint identification process.

The NIST engineers first studied the manual methods used by human fingerprint technicians to make identifications. These methods were based on comparing the minutiae (i.e., ridge endings and ridge bifurcations) on fingerprint ridges. If the minutiae from two fingerprints were determined to be topologically equivalent, the two fingerprints were declared to be identical—that is, having been recorded from the same finger of the same person. After this review, and after studying additional problems inherent with the inking process, they believed that a computerized solution to automatically match and pair minutiae could be developed that would operate in a manner similar to the techniques used by human examiners to make fingerprint identifications. But to achieve this goal, three major tasks would have to be accomplished. First, a scanner had to be developed that could automatically read and electronically capture the inked fingerprint image. Second, it was necessary to accurately and consistently detect and identify minutiae existing in the captured image. Finally, a method had to be developed to compare two lists of minutiae descriptors to determine whether they both most likely came from the same finger of the same individual.

The Identification Division of the FBI decided that the approach suggested by Moore and Wegstein should be followed. To address the first two of the three tasks, on December 16, 1966, the FBI issued a Request for Quotation (RFQ) "for developing, demonstrating, and testing a device for reading certain fingerprint minutiae" (FBI, 1966). This contract was for a device to automatically locate and determine the relative position and orientation of the specified minutiae in individual fingerprints on standard fingerprint cards to be used for testing by the FBI. The requirements stated that the reader must be able to measure and locate minutiae in units of not more than 0.1 mm and that the direction of each minutiae must be measured and presented as output in units of not more than 11.25 degrees (1/32 of a full circle). The initial requirements called for a prototype model to process 10,000 single fingerprints (1,000 cards). Contractors were also instructed to develop a proposal for a subsequent contract to process 10 times that number of fingerprints.

The 14 proposals received in response to this RFQ were divided into 5 broad technical approaches. At the conclusion of the proposal evaluation, two separate proposals were funded to provide a basic model for reading fingerprint images and extracting minutiae. Both proposed to use a "flying spot scanner" for capturing the image. But each offered a different approach for processing the captured image data, and both seemed promising. One contract was awarded to Cornell Aeronautical Labs, Inc., which proposed using a general-purpose digital computer to process binary pixels and develop programs for detecting and providing measurement parameters for each identified minutiae. The second contract was awarded to North American Aviation, Inc., Autonetics Division, which proposed using a special-purpose digital process to compare fixed logical marks to the image for identifying, detecting, and encoding each minutia.

While the devices for fingerprint scanning and minutiae detection were being developed, the third task of comparing two minutiae lists to determine a candidate match was addressed by Joe Wegstein (Wegstein, 1969a, 1970, 1972a/b, 1982; Wegstein and Rafferty, 1978, 1979; Wegstein et al., 1968). He developed the initial algorithms for determining fingerprint matches based on the processing and comparison of two lists describing minutiae location and orientation. For the next 15 years, he continued to develop more reliable fingerprint matching software that became increasingly more complex in order to account for such things as plastic distortion and skin elasticity. Algorithms he developed were embedded in AFISs that were eventually placed in operation at the FBI and other law enforcement agencies.

By 1969, both Autonetics and Cornell had made significant progress on their feasibility demonstration models. In 1970, a Request for Proposal (RFP) was issued for the construction of a prototype fingerprint reader to reflect the experience gained from the original demonstration models with an additional requirement for speed and accuracy. Cornell was awarded the contract to deliver the prototype reader to the FBI in 1972. After a year's experience with the prototype system, the FBI issued a new RFP containing additional requirements such as a high-speed card-handling subsystem. In 1974, Rockwell International, Inc., was awarded a contract to build five production model automatic fingerprint reader systems. This revolutionary system was called Finder. These readers were delivered to the FBI in 1975 and 1976. The next 3 years were devoted to using these readers in the conversion of 15 million criminal fingerprint cards (Moore, 1991, pp 164–175).

As it became apparent that the FBI's efforts to automate the fingerprint matching process would be successful, state and local law enforcement agencies began to evaluate this new technology for their own applications. The Minneapolis– St. Paul system in Minnesota was one of the first automated fingerprint matching systems (after the FBI's) to be installed in the United States. Further, while the United States was developing its AFIS technology in the 1960s, France, the United Kingdom, and Japan were also doing research into automatic fingerprint image processing and matching.

## 1.4. French AFIS Initiative

In 1969, M. R. Thiebault, Prefecture of Police in Paris, reported on the French efforts. (Descriptions of work done by Thiebault can be found in the entries listed in the Additional Information section of this chapter.) France's focus was on the solution to the latent fingerprint problem rather than the general identification problem that was the concern in the United States. The French approach incorporated a *vidicon* (a video camera tube) to scan photographic film transparencies of fingerprints. Scanning was done at 400 pixels per inch (ppi), which was less than an optimal scan rate for latent work. This minutiae matching approach was based on special-purpose, high-speed hardware that used an array of logical circuits. The French also were interested in resolving the problem of poor fingerprint image quality. In order to acquire a high-contrast image that would be easy to photograph and process, a technique was developed to record live fingerprint images photographically using a principle of "frustrated total internal reflection" (FTIR). Although not put into large-scale production at that time, 20 years later FTIR became the cornerstone for the development of the modern-day livescan fingerprint scanners. These are making the use of ink and cards obsolete for nonforensic identification purposes today.

By the early 1970s, the personnel responsible for development of France's fingerprint automation technology had changed. As a result, there was little interest in pursuing automated fingerprint identification research for the next several years. In the late 1970s, a computer engineering subsidiary of France's largest financial institution responded to a request by the French Ministry of Interior to work on automated fingerprint processing for the French National Police. Later, this company joined with the Morphologic Mathematics Laboratory at the Paris School of Mines to form a subsidiary called Morpho Systems that went on to develop a functioning. Currently, Morpho Systems is part of Sagem (also known as Group SAFRAN).

## 1.5. United Kingdom AFIS Initiative

During the same period of time, the United Kingdom's Home Office was doing research into automatic fingerprint identification. Two of the main individuals responsible for the United Kingdom's AFIS were Dr. Barry Blain and Ken Millard. (Papers produced by Millard are listed in the Additional Information section of this chapter). Like the French, their main focus was latent print work. By 1974, research was being done in-house with contractor assistance from Ferranti, Ltd. The Home Office developed a reader to detect minutiae, record position and orientation, and determine ridge counts to the five nearest neighbors to the right of each minutia. This was the first use of ridge count information by an AFIS vendor (Moore, 1991).

## 1.6. Japanese AFIS Initiative

Like France and the United Kingdom, Japan's motivation for a fingerprint identification system was directed toward the matching of latent images against a master file fingerprints. Japan's researchers believed that an accurate latent system would naturally lead to the development of an accurate tenprint system.

By 1966, the Osaka Prefecture Police department housed almost 4 million single fingerprints. An early automation effort by this agency was the development of a pattern classification matching system based on a 17- to 20-digit number encoded manually (Kiji, 2002, p 9). Although this approach improved the efficiency of the totally manual method enormously, it had inherent problems. It required a great deal of human precision and time to classify the latents and single fingerprints; was not fully suitable for latent matching; and produced a long list of candidates, resulting in expensive verifications.

Within a few years, the fingerprint automation focus of Japanese researchers had changed. By 1969, the Identification Section of the Criminal Investigation Bureau, National Police Agency of Japan (NPA), approached NEC to develop a system for the computerization of fingerprint identification. NEC determined that it could build an automated fingerprint identification system employing a similar minutiae-based approach to that being used in the FBI system under development. At that time, it was thought that a fully automated system for searching fingerprints would not be realized for 5 to 10 years. In 1969, NEC and NPA representatives visited the FBI and began to learn about the current state of the art for the FBI's AFIS plans. During the same period, NPA representatives also collaborated with Moore and Wegstein from NIST. Additional AFIS sites were visited where information was acquired regarding useful and worthless approaches that had been attempted. All of this information was evaluated and used in the development of the NEC system.

For the next 10 years, NEC worked to develop its AFIS. In addition to minutiae location and orientation, this system also incorporated ridge-count information present in the local four surrounding quadrants of each minutiae under consideration for pairing. By 1982, NEC had successfully installed its system in the NPA and started the card conversion process. Within a year, latent inquiry searches began.

In 1980, NEC received a U.S. patent for automatic minutiae detection. It began marketing its automated fingerprint identification systems to the United States a few years later.

## 1.7. The Politicization of Fingerprints and the San Francisco Experiment

Early development and implementation of automated fingerprint systems was limited to national police agencies in Europe, North America, and Japan. But the problems associated with huge national databases and the newborn status of computer technology in the 1970s limited the utility of these systems. Government investment in AFIS was justified largely on the promise of efficiency in the processing of incoming tenprint records. But funding these expensive systems on the local level would demand some creativity (Wayman, 2004, pp 50–52).

Following the success of the FBI's Finder, Rockwell took its system to market in the mid-1970s. Rockwell organized a users group for its Printrak system and sponsored an annual conference for customers and would-be customers. Starting with a beta-site in San Jose, California, more than a dozen installations were completed in quick succession. Peggy James of the Houston Police Department, Joe Corcoran from Saint Paul, Donna Jewett from San Jose, and others devoted their energies toward educating the international fingerprint community on the miracle of the minutiae-based Printrak system. Each system that came online trumpeted the solution of otherwise unsolvable crimes and the identity of arrested criminals. A users group newsletter was published and distributed that highlighted some of the best cases and listed the search statistics of member agencies.

Ken Moses of the San Francisco Police Department had attended several of those Printrak conferences and became a staunch crusader for fingerprint automation. In three successive years, he persuaded the Chief of Police to include a Printrak system in the city budget, but each time it was vetoed by the mayor. After the third mayoral veto, a ballot proposition was organized by other politicians. The proposition asked citizens to vote on whether they wanted an automated fingerprint system. In 1982, Proposition E passed with an 80% plurality.

The mayor refused to approve a sole-source purchase from Rockwell, even though it was the only system in the world being marketed. She insisted on a competitive bid with strict evaluation criteria and testing. While on a trade mission to Japan, the mayor learned that the Japanese National Police were working with NEC to install a fingerprint system, but NEC stated that the system was being developed as a public service and the company had no plans to market it. After meeting with key Japanese officials, NEC changed its mind and agreed to bid on the San Francisco AFIS.

When the bids were opened, not only had Printrak and NEC submitted proposals, but a dark horse named Logica had also entered the fray. Logica had been working with the British Home Office to develop a system for New Scotland Yard.

San Francisco retained systems consultant Tim Ruggles to assist in constructing the first head-to-head benchmark tests of competing in-use fingerprint systems. The test was most heavily weighted toward latent print accuracy, and a set of 50 latent prints graded from poor to good from actual past cases was searched against a prescribed ten-print database. All tests were conducted at the respective vendor's home site.[2] NEC was awarded the contract and installation was completed in December 1983.

Besides being the first competitive bid on 1980s technology, what differentiated the San Francisco system from those that had gone before was organizational design. AFIS was viewed as a true system encompassing all aspects of friction ridge identification—from the crime scene to the courtroom. The AFIS budget included laboratory and crime scene equipment, training in all phases of forensic evidence, and even the purchase of vehicles. In 1983, a new crime scene unit was organized specifically with the new system as its centerpiece. Significant organizational changes were put into effect:

1. All latents that met minimum criteria would be searched in AFIS.
2. A new unit called Crime Scene Investigations was created and staffed on a 24/7 schedule.
3. Department policies were changed to mandate that patrol officers notify crime scene investigators of all felonies with a potential for latent prints.
4. All crime scene investigators who processed the crime scenes were trained in the use of the system and encouraged to search their own cases.
5. Performance statistics were kept from the beginning, and AFIS cases were tracked through the criminal justice system to the courts.

The result of the San Francisco experiment was a dramatic 10-fold increase in latent print identifications in 1984. The district attorney demanded and got five new positions to prosecute the AFIS cases. The conviction rate in AFISgenerated burglary cases was three times higher than in burglary cases without this type of evidence (Figure 1; Bruton, 1989).

At a time when burglary rates were steeply rising in cities across the nation, the burglary rate plummeted in San Francisco (Figure 2; Bruton, 1989). Reporters, academics, and police administrators from around the world inundated the San Francisco Police Department for demonstrations and information.

The importance of politics and publicity was not lost on other agencies. Los Angeles even enlisted the backing of film stars to stir up public support. The identification of serial killer Richard Ramirez, the infamous Night Stalker, through a search of the brand-new California State AFIS made worldwide headlines and guaranteed the future funding of systems in California.

## 1.8. AFIS Proliferation

The widely publicized success in San Francisco provided the spark for the rapid proliferation of new AFIS installations along with a methodology of benchmark testing to evaluate the claims of the growing number of competing vendors. Governments quickly provided funding so that, by 1999, the International Association for Identification's (IAI's) *AFIS Directory of Users* identified 500 AFIS sites worldwide (IAI, 1999).

The burgeoning market in these multimillion-dollar systems put forensic identification on the economic map. Commercial exhibits at IAI's conferences that had formerly featured companies hawking tape and powder now expanded to digital image enhancement, lasers and forensic light sources, and the latest in new developments from Silicon Valley. The San Francisco Crime Lab received its first digital imaging system in 1986. This 3M/Comtal system was dedicated to friction ridge enhancement. Fingermatrix installed the first livescan device in the San Francisco Police Identification Bureau in 1988. AFIS brought crime scene and forensic identification out of the basement; no local or state law enforcement administrator wanted to be accused of being left behind.

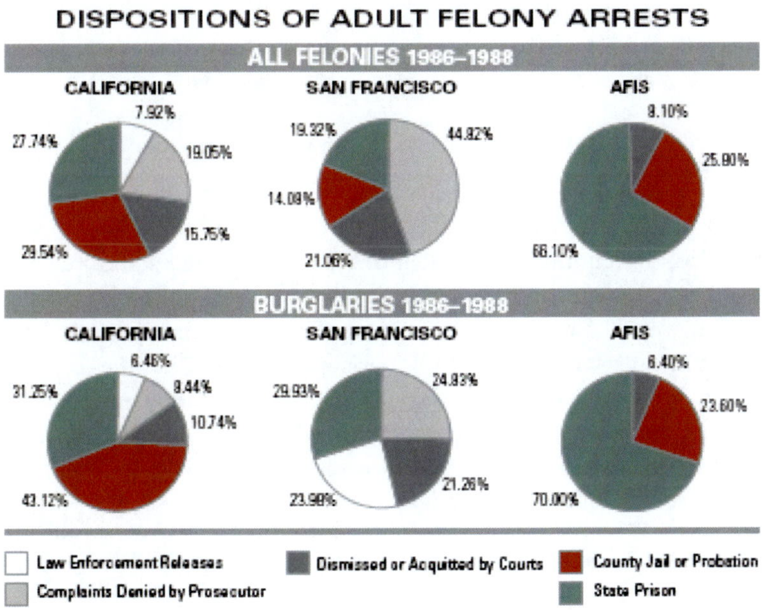


Figure 1. Tracking latent hits through the courts. (Bruton, 1989).

However, the frenzied expansion of AFIS was not always logical and rational. By the early 1990s, the four biggest vendors—Printrak, NEC, Morpho, and Cogent—were in competition, each offering proprietary software that was incompatible with the others, especially in latent print searching.

Expansion was often based on political considerations and competing mission priorities. Local and state agencies expressed differences in priorities in terms of system design, with states generally emphasizing criminal identification or tenprint functions, while cities and counties focused on crime solving or latent print functions.

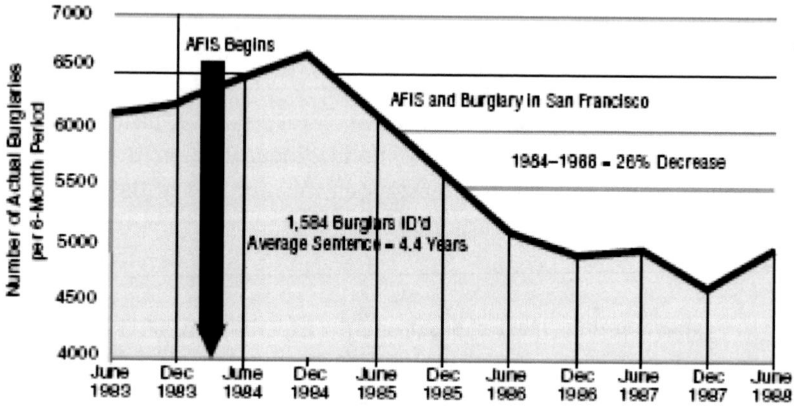


Figure 2. Statistical study of AFIS hits vs. burglaries in San Francisco, 1984–1988. (Bruton, 1989).

Generally, the demands of latent print processing on computer resources far exceeded the requirements of tenprint processing, and states balked at the additional expense and technical complexity. As a result, cities, counties, and states often went their separate ways, installing dissimilar systems that could not communicate with neighboring jurisdictions or with the central state repository. Vendors eagerly encouraged this fragmentation in an attempt to gain market share and displace competitors whenever possible. The evolution of electronic transmission standards (see section 3) ameliorated this problem for tenprint search but not for latent search.

# 2. AFIS Operations

## 2.1. AFIS Functions and Capabilities

Identification bureaus are legislatively mandated to maintain criminal history records. Historically, this meant huge file storage requirements and cadres of clerks to maintain and search them. Demographic-based criminal history computers were established well ahead of AFIS, first as IBM card sort systems and then as all-digital information systems with terminals throughout the state and, via the National Crime Information Center (NCIC) network and the National Law Enforcement Teletype System (Nlets), throughout the nation. These automated criminal history systems became even more labor-intensive than the paper record systems they supposedly replaced. In many systems, more paper was generated and placed into the history jackets along with the fingerprint cards, mug shots, warrants, and other required documents.

AFIS revolutionized state identification bureaus because it removed from the paper files the last document type that could not previously be digitized—the fingerprint card. State identification bureaus could now bring to their legislatures cost–benefit analyses that easily justified the purchase of an automated fingerprint system through the reduction of clerical personnel.

Local and county jurisdictions did not usually enjoy the economic benefits of state systems. Pre-AFIS personnel levels were often lower and controlled more by the demands of the booking process than by file maintenance. AFIS generally increased staffing demands on the latent and crime-scene-processing side because it made crime scene processing dramatically more productive. Local and county AFIS purchases were usually justified on the basis of their crime-solving potential.

### *2.1.1. Technical Functions*

Law enforcement AFISs are composed of two interdependent subsystems: the tenprint (i.e., criminal identification) subsystem and the latent (i.e., criminal investigation) subsystem. Each subsystem operates with a considerable amount of autonomy, and both are vital to public safety.

The tenprint subsystem is tasked with identifying sets of inked or livescan fingerprints incident to an arrest or citation or as part of an application process to determine whether a person has an existing record.

In many systems, identification personnel are also charged with maintaining the integrity of the fingerprint and criminal history databases. Identification bureau staffs are generally composed of fingerprint technicians and supporting clerical personnel.

An automated tenprint inquiry normally requires a minutiae search of only the thumbs or index fingers. Submitted fingerprints commonly have sufficient clarity and detail to make searching of more than two fingers unnecessary. Today's AFIS can often return a search of a million records in under a minute. As databases have expanded across the world, some AFIS engineers have expanded to searching four fingers or more in an effort to increase accuracy.

The latent print or criminal identification subsystem is tasked with solving crimes though the identification of latent prints developed from crime scenes and physical evidence. Terminals used within the latent subsystem are often specialized to accommodate the capture and digital enhancement of individual latent prints. The latent subsystem may be staffed by latent print examiners, crime scene investigators, or laboratory or clerical personnel. The staff of the latent subsystem is frequently under a different command structure than the tenprint subsystem and is often associated with the crime laboratory.

The search of a latent print is more tedious and time- consuming than a tenprint search. Latent prints are often fragmentary and of poor image quality. Minutiae features are normally reviewed one-by-one before the search begins. Depending on the portion of the database selected to be searched and the system's search load, the response may take from a few minutes to several hours to return.

Most law enforcement AFIS installations have the ability to perform the following functions:

- Search a set of known fingerprints (tenprints) against an existing tenprint database (TP–TP) and return with results that are better than 99% accurate.[3]
- Search a latent print from a crime scene or evidence against a tenprint database (LP–TP).
- Search a latent from a crime scene against latents on file from other crime scenes (LP–LP).
- Search a new tenprint addition to the database against all unsolved latent prints in file (TP–LP).

Enhancements have been developed to allow other functions that expand AFIS capabilities, including:

- Addition of palmprint records to the database to allow the search of latent palmprints from crime scenes.
- Interfacing of AFIS with other criminal justice information systems for added efficiency and "lights out"[4] operation.
- Interfacing of AFIS with digital mug shot systems and livescan fingerprint capture devices.
- Addition of hand-held portable devices for use in identity queries from the field. The query is initiated by scanning one or more of the subject's fingers, extracting the minutiae within the device, and transmitting to AFIS, which then returns a hit or no-hit (red light, green light) result. Hit notification may be accompanied by the thumbnail image of the subject's mug shot.
- Multimodal identification systems, including fingerprint, palmprint, iris, and facial recognition, are now available.

## 2.2. System Accuracy

Most dedicated government computer systems are based on demographic data such as name, address, date of birth, and other information derived from letters and numbers. For example, to search for a record within the motor vehicle database, one would enter a license number or operator data. The success of the search will be dependent on the accuracy with which the letters and numbers were originally perceived and entered. The inquiry is straightforward and highly accurate at finding the desired record.

Automated fingerprint systems are based on data extracted from images. Although there is only one correct spelling for a name in a motor vehicle database, a fingerprint image can be scanned in an almost infinite number of ways. Success in searching fingerprints depends on the clarity of the images and the degree of correspondence between the search print and the database print (compression and algorithms are two other factors that can affect accuracy). In the case of searching a new tenprint card against the tenprint database, there is usually more than enough image information present to find its mate 99.9% of the time in systems with operators on hand to check respondent lists (rather than true "lights out" operations).

A latent print usually consists of a fragmentary portion of a single finger or piece of palm, though the quality of some latent impressions can exceed their corresponding images of record. The amount of information present in the image is usually of lesser quality and often is contaminated with background interference. Entering latents into the computer has a subjective element that is based on the experience of the operator. Based on latent print acceptance test requirements commonly found in AFIS proposals and contracts, the chances of a latent print finding its mate in the database is about 70 to 80%. Naturally, the better the latent image, the higher the chances of success. Inversely, the chance of missing an identification, even when the mate is in the database, is 25%. Especially in latent print searches, failure to produce an identification or a hit does not mean the subject is not in the database. Other factors beyond the knowledge and control of the operator, such as poor-quality database prints, will adversely affect the chances of a match.

Because of the variability of the images and the subjectivity of the terminal's operator, success is often improved by conducting multiple searches while varying the image, changing operators, or searching other systems that may contain different copies of the subject's prints. It is common that success comes only on multiple attempts.

## 2.3. Peripheral Benefits

### *2.3.1. Community Safety*

There is no national reporting mechanism for the gathering of AFIS (or latent print) statistics, so the measurable benefits are illusive. However, to provide some recognition of those benefits, the author of this chapter conducted a survey of latent hits in the 10 largest states by population for the year 2005 (Table 1). Prior attempts to provide this type of information have revealed inconsistencies in how identifications are counted and how the hit rate is determined (Komarinski, 2005, pp 184–189).

Based on the author's survey, an estimated 50,000 suspects a year in the United States are identified through AFIS latent searches. In conducting the survey, if the contacted state bureaus did not have statewide figures, attempts were made to also contact the five largest cities in that state. (In no instance was it possible to contact every AFIS-equipped jurisdiction in a state, so the total hits are the minimum number of hits.) Also, only case hits or suspect hits were counted, depending on what data each agency kept. (When agencies reported multiple hits to a single person, this was not included in data presented.)

Extrapolating from the table, if the remaining 40 states and all agencies of the federal government each had just one latent hit per day, the total estimate of latent hits for the entire United States would surpass 50,000.

**Table 1. Minimum hits (cases or persons identified) from 10 largest states by population for 2005**

| | | |
|---|---|---|
| 1 | California | 8,814 |
| | | |
| 3 | New York | 2,592 |
| | | |
| 5 | Illinois | 1,224 |
| | | |
| 7 | Ohio* | 1,495 |
| | | |
| 9 | Georgia | 980 |
| | | |
| | Total | 29,178 |

* Cleveland not available.

** Detroit not available.

Few studies have been done to measure what effect, if any, a dramatic increase in the rate of suspect latent print identifications from AFIS has had on public safety overall. The burglary data from San Francisco in the late 1980s (Figure 2) is probative but must be narrowly construed. FBI Uniform Crime Reports show a steady decline in most serious offenses that coincide with the proliferation of AFIS, but no cause-and-effect relationship has been explored by academia or government. During the 1990s, many states passed "three strikes" laws increasing the punishment for felony offenses that some theorists have held are responsible for the decline in crime. But before harsher penalties can be applied, perpetrators must be identified and apprehended.

Burglary is the offense most impacted by AFIS. Assume that an active burglar is committing two offenses per week when he is apprehended on the basis of an AFIS hit. He is convicted and, based on harsh sentencing laws, sent to prison for 5 years. In this case, that one AFIS hit will have prevented 100 crimes per year over the course of the 5 year sentence. If this one arrest is then multiplied by some fraction of the totals from the table above, a truer appreciation of the impact that AFIS is having on society can be gained.

### *2.3.2 .Validation of Friction Ridge Science*

There are many ways to test the efficacy of a theoretical proposition. Corporate and academic laboratories pour tremendous resources into building models that they hope will closely duplicate performance in the real world. Even after successfully passing such testing, theories fail and products get recalled after weathering the rigors of the real world. In-use models invariably trump laboratory models.

During the past 100 years, many models have been constructed to test the theory that no two friction ridge images from different areas of palmar surfaces are alike and to determine what minimum number of minutiae is sufficient to support an individualization decision.

Automated fingerprint systems have been effectively testing identification theory millions of times a day every day for more than 20 years. These systems tend to validate what friction ridge examiners have propounded since Galton first set forth his standards. AFIS has also served as a catalyst to help examiners expand their image-processing knowledge and skills.

Some errors occur every year in both manual and automated systems, and it is through the study of errors that both systems can be improved in the future. According to Dr. James Wayman, Director of the National Biometrics Test Center, "Error rates (in friction ridge identification) are difficult to measure, precisely because they are so low" (Wayman, 2000)

## 2.4. IAFIS

The Integrated Automated Fingerprint Identification System, more commonly known as IAFIS, is the world's largest collection of criminal history information. Fully operational since July 28, 1999, IAFIS is maintained by the FBI's Criminal Justice Information Services (CJIS) Division in Clarksburg, WV, and contains fingerprint images for more than 64 million individuals. The FBI's CJIS Division system's architecture and the identification and investigative services provided by the division form an integrated system-of-services (SoS) concept. These identification and information services enable local, state, federal, tribal, and international law enforcement communities, as well as civil organizations, to efficiently access or exchange critical information 24 hours a day, 365 days per year. The SoS provides advanced identification and ancillary criminal justice technologies used in the identification of subjects.

The systems within the CJIS SoS, including IAFIS, have evolved over time, both individually and collectively, to add new technological capabilities, embrace legislative directives, and improve the performance and accuracy of their information services. During its first year of inception, IAFIS processed nearly 14.5 million fingerprint submissions. Today, IAFIS processes similar tenprint volumes in as little as 3 to 4 months. Although designed to respond to electronic criminal transactions within 2 hours and civil transactions within 24 hours, IAFIS has exceeded these demands, often providing criminal search requests in less than 20 minutes and civil background checks in less than 3 hours. Likewise, IAFIS provides the latent print examiners with a superlative investigative tool, allowing fingerprint evidence from crime scenes to be searched in approximately 2 hours rather than the 24-hour targeted response time. Although declared a successful system early within its deployment, IAFIS continues to improve as a vital asset to law enforcement agencies more than 10 years later. Today's transient society magnifies the need for an economic, rapid, positive identification process for both criminal and noncriminal justice background checks. IAFIS processes are regularly improved to allow for a quick and accurate fingerprint-based records check, whether related to terrorists trying to enter the United States or applicants seeking positions of trust. Figure 3 illustrates the states that currently interface with IAFIS electronically.

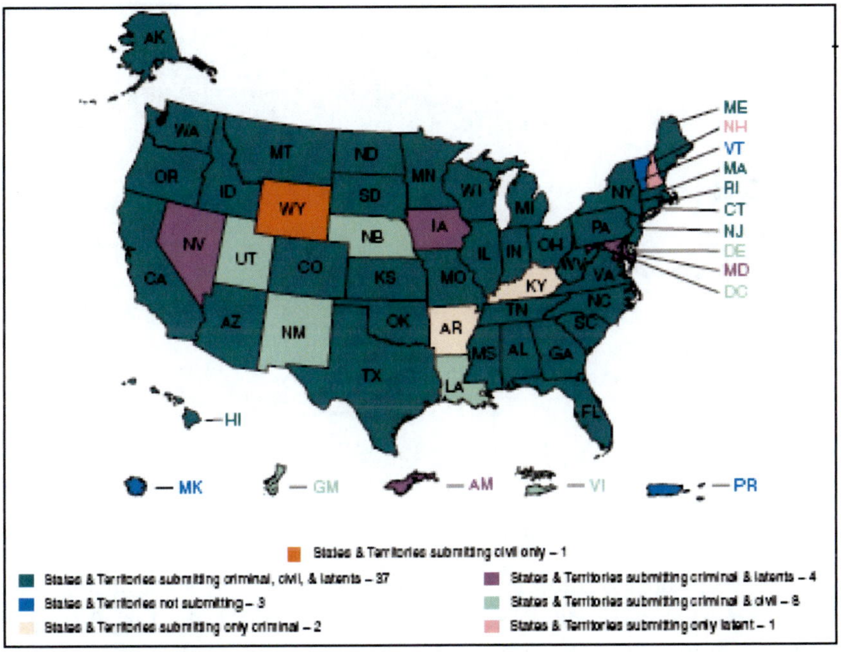


Figure 3. Electronic submissions to IAFIS. (Illustration from the Federal Bureau of Investigation).

The increasingly complex requirements of the SoS architecture demand a well-structured process for its operations and maintenance. Each of these systems has multiple segments consisting of computer hardware and software that provide the operating systems and utilities, database management, workflow management, transaction or messaging management, internal and external networking, communications load balancing, and system security. IAFIS consists of three integrated segments: the Identification Tasking and Networking (ITN) segment, the Interstate Identification Index (III), and AFIS (Figure 4).

Within IAFIS, the ITN segment acts as a "traffic cop" for the fingerprint system, providing workflow/workload management for tenprint, latent print, and document processing. The ITN provides the human–machine interfaces, the internal interfaces for communications within the IAFIS backbone communications element, the storage and retrieval of fingerprint images, the external communications interfaces, the IAFIS back-end communications element, and user fee billing.

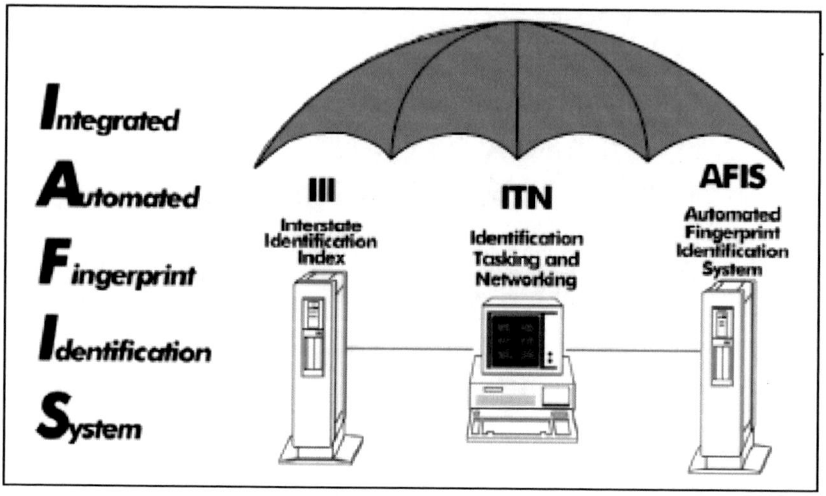


Figure 4. IAFIS segments. (Illustration from the Federal Bureau of Investigation).

The III provides subject search, computerized criminal history, and criminal photo storage and retrieval. The AFIS searches the FBI fingerprint repository for matches to tenprint and latent fingerprints. Supporting IAFIS is the CJIS-wide area network (WAN), providing the communications infrastructure for the secure exchange of fingerprint information to and from external systems. The external systems are the state control terminal agencies, state identification bureaus, and federal service coordinators.

Also submitting fingerprint information to IAFIS is the Card Scanning Service (CSS). The CSS acts as a conduit for agencies that are not yet submitting fingerprints electronically. The CSS makes the conversion of fingerprint information from paper format to electronic format and submits that information to IAFIS. Another system providing external communications for IAFIS is Nlets. The purpose of Nlets is to provide interstate communications to law enforcement, criminal justice, and other agencies involved in the enforcement of laws. Figure 5 depicts the high-level IAFIS architecture. Users wishing to interface with IAFIS electronically must comply with the FBI's Electronic Fingerprint Transmission Specification (EFTS).

Electronic access to and exchange of fingerprint information with the world's largest national repository of automated criminal and civil records is fulfilling the CJIS mission:

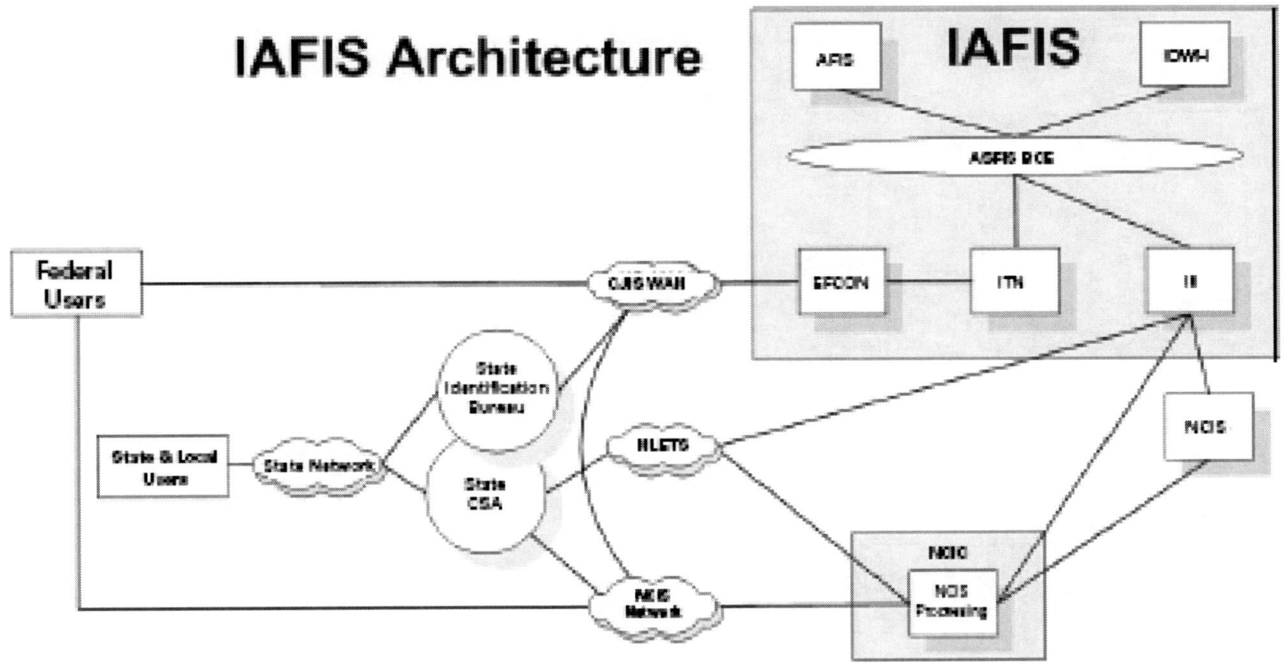


Figure 5. IAFIS networked architecture. (Illustration from the Federal Bureau of Investigation.).

> The CJIS Division mission is to reduce terrorist activities by maximizing the ability to provide timely and relevant criminal justice information to the FBI and to qualified law enforcement, criminal justice, civilian, academic, employment, and licensing agencies concerning individuals, stolen property, criminal organizations and activities, and other law enforcement-related data.

### *2.4.1. IAFIS Status As of Early 2006*

Because of the evolutionary changes to the American National Standards Institute (ANSI)/NIST standard in 1997, 2000, and 2006, the FBI has not always had the financial resources or corporate commitment to update IAFIS and keep it current. One area where it has moved forward is the acceptance and processing of "segmented slaps" for civil transactions. These transactions use a modified livescan platen that is 3 inches high so the four fingers of each hand can be placed as a "slap" in a straight up-and-down position. Similarly, both thumbs can be captured simultaneously for a total of three images (type 4 or type 14 as defined in sections 3.2.1 and 3.3). The resultant transaction's three-image files are easy to segment with the capture device software. The three images and relative location of the segmented fingers within the images are all transmitted. This dramatically reduces collection time and improves the captured-image quality from a content perspective due to the flat, straight, 3-inch placement.

One drawback to IAFIS is that it cannot store and search palmprints, though several production AFISs can do so. Also, at least one foreign production and several domestic AFIS sites accept and store 1,000-pixels-per-inch tenprint images—IAFIS cannot yet do this.

The FBI recognizes its need to expand its services and has (1) tested small palm systems and (2) started a project known as the Next Generation Identification Program (NGI). Driven by advances in technology, customer requirements, and growing demand for IAFIS services, this program will further advance the FBI's biometric identification services, providing an incremental replacement of current IAFIS technical capabilities while introducing new functionality. NGI improvements and new capabilities will be introduced across a multiyear time frame within a phased approach. The NGI system will offer state-of-the-art biometric identification services and provide a flexible framework of core capabilities that will serve as a platform for multimodal functionality.

#### *2.4.2. Universal Latent Work Station*

AFISs that are fully ANSI/NIST compliant can send image-based transactions from site to site. But in the latent community, most practitioners want to edit the images and extract the minutiae themselves, that is, perform remote searches rather than submittals. This model also plays well with the ability of most agencies to provide the skilled labor required for imaged-based submittals from other agencies.

The FBI CJIS Division addressed this issue by working closely with Mitretek and the four major AFIS vendors to develop a set of tools that would permit the creation of remote searches for any of their automated fingerprint identification systems and for IAFIS. The result is a free software product called the Universal Latent Workstation (ULW). This software can run on a stand-alone PC with either a flatbed scanner or a digital camera interface. It can also run on vendor-provided latent workstations. At a minimum, when specifying an AFIS in a procurement, one should mandate that the AFIS be able to generate remote searches to IAFIS. It is further recommended that the procurer ask for the ability to perform the ULW function so the vendors can integrate ULW into their systems.

The ULW also provides the ability to launch latent print image searches into IAFIS without the need to manually encode minutiae when working with high-quality latent prints.

# 3. STANDARDS

## 3.1. Background

Standards are mutually agreed upon attributes of products, systems, communication protocols, and so forth. Standards are what permit people to purchase light bulbs made in Hungary, the United States, or Japan and know they will fit in a standard lamp socket. Industries and governments establish standards not just for the convenience of the consumer but to permit competition for the same product.

Each nation has its own standards bureau or management body. In the United States, it is ANSI. At the international level, there are several such bodies. They include the United Nation's International Labor Organization (ILO) and International Civil Aviation Organization (ICAO), the International Criminal Police Organization (Interpol), the International Standards Organization (ISO), and the International Electrotechnical Commission (IEC).

Other than the United Nations and Interpol, these standards bodies do not "invent" or "create" standards but rather provide processes that authorized bodies can use to propose standards for approval at the national level and then at the international level. The United Nations and Interpol tend to build on these national and international standards bodies' standards rather than starting from scratch.

ANSI has offices in both New York and Washington, DC. ANSI has authorized more than 200 bodies to propose standards. If all the procedures are followed correctly and there are no unaddressed objections, then the results of the efforts of these bodies become ANSI standards. The 200 organizations include the following:

- The Department of Commerce's NIST
- IAI
- The American Association of Motor Vehicle Administrators
- The International Committee for Information Technology Standards (INCITS)

## 3.2. Fingerprint Standards

Law enforcement agencies around the world have had standards for the local exchange of inked fingerprints for decades. In 1995, Interpol held a meeting to address the transfer of ink-and-paper fingerprint cards (also known as forms) between countries. The local standards naturally had different text fields, had different layouts of text fields, were in different languages, and were on many different sizes of paper. Before that effort could lead to an internationally accepted fingerprint form, Interpol moved to the electronic exchange of fingerprints.

In the ink-and-paper era, the standards included fiber content and thickness of the paper, durability of the ink, size of the "finger boxes", and so forth. With the move in the early 1990s toward near real-time responses to criminal fingerprint submittals, there came a new set of standards.

The only way to submit, search, and determine the status of fingerprints in a few hours from a remote site is through electronic submittal and electronic responses. The source can still be ink-and-paper, but the images need to be digitized and submitted electronically to address the growing demand for rapid turnaround of fingerprint transactions.

The FBI was the first agency to move to large-scale electronic submission of fingerprints from remote sites. As part of the development of IAFIS, the FBI worked very closely with NIST to develop appropriate standards for the electronic transmission of fingerprint images.

Starting in 1991, NIST held a series of workshops with forensic experts, fingerprint repository managers, industry representatives, and consultants to develop a standard, under the ANSI guidelines, for the exchange of fingerprint images. It was approved in November 1993, and the formal title was "Data Format for the Interchange of Fingerprint Information (ANSI NIST-CSL 1-1993)". This standard was based on the 1986 ANSI/National Bureau of Standards minutiae-based standard and ANSI/NBS-ICST 1-1986, a standard that did not address image files.

This 1993 NIST standard (and the later revisions) became known in the fingerprint technology world simply as the "ANSI/NIST standard". If implemented correctly (i.e., in full compliance with the standard and the FBI's implementation), it would permit fingerprints collected on a compliant livescan from any vendor to be read by any other compliant AFIS and the FBI's yet-to-be-built (at that time) IAFIS.

The standard was deliberately open to permit communities of users (also known as domains of interest) to customize it to meet their needs. Some of the customizable areas were image density (8-bit gray scale or binary) and text fields associated with a transaction (e.g., name, crime). The idea was that different communities of users would write their own implementation plans. The mandatory parts of the ANSI/NIST standard were the definitions of the record types, the binary formats for fingerprint and signature images and, within certain record types, the definition of "header" fields such as image compression type.

### *3.2.1. Record Types*

For a transaction to be considered ANSI/NIST compliant, the data must be sent in a structured fashion with a series of records that align with ANSI/NIST record types as implemented in a specific user domain (e.g., Interpol).

- All transmissions (also known as transactions) have to start with a type 1 record that is basically a table of contents for the transmission, the transaction type field (e.g., CAR for "criminal tenprint submission—answer required"), and the identity of both the sending and receiving agencies.

- Type 2 records can contain user-defined information associated with the subject of the fingerprint transmission (such as name, date of birth, etc.) and the purpose of the transaction (arrest cycle, applicant background check, etc.). These fields are defined in the domain-of-interest implementation standard (e.g., the FBI's EFTS). Note that type 2 records are also used for responses from AFISs. They fall into two sets: error messages and search results. The actual use is defined in the domain specification.
- Types 3 (low-resolution gray scale), 4 (high-resolution gray scale), 5 (low-resolution binary), and 6 (high-resolution binary) were set up for the transmission of fingerprint images at different standards (500 ppi for high resolution and 256 ppi for low resolution) and image density (8 bits per pixel for grayscale) or binary (1 bit per pixel for black and white). Note that all images for records type 3 through 6 are to be acquired at a minimum of 500 ppi; however, low-resolution images are down-sampled to 256 ppi for transmission. There are few, if any, ANSI/NIST implementations that support type 3, 5, or 6 images (see explanation below). None of these three record types are recommended for use by latent examiners and fingerprint technicians.
- Type 7 was established for user-defined images (e.g., latent images, faces) and, until the update of the ANSI/ NIST standard in 2000, it was the record type for exchanging latent images. This record type can be used to send scanned copies of identity documents, and so forth. Again, the domain specification determines the legitimate uses of the type 7 record.
- Type 8 was defined for signatures (of the subject or person taking the fingerprints), and it is not used in many domains.
- Type 9 was defined for a minimal set of minutiae that could be sent to any AFIS that was ANSI/NIST-compliant.

The first such implementation plan was the FBI's EFTS issued in 1994. The EFTS limited what record types, of the nine defined in the ANSI/NIST standard, the FBI would use, and defined the type 2 data fields. The key decision the FBI made was that it would only accept 500-ppi gray-scale images or, in ANSI/NIST parlance, type 4 images. As a result of that decision, all law enforcement systems since then have specified type 4 images and do not accept types 3, 5, or 6, which as a result have fallen into disuse for these applications in the United States.

The type 4 records start out with header information in front of the image. The headers tell the computer which fi nger the image is from, whether it is from a livescan or an inked card, the image size in the number of pixels of width and height, and whether the image is from a rolled impression or a flat or plain impression.

### *3.2.2. Image Quality*

Both the ANSI/NIST standard and the EFTS lacked any metrics or standards for image quality. The FBI then appended the EFTS with an image quality standard (IQS) known as Appendix F. (Later, a reduced set of image quality specifications were added as Appendix G because the industry was not uniformly ready to meet Appendix F standards.) The IQS defines minimal acceptable standards for the equipment used to capture the fi ngerprints. There are six engineering terms specified in the IQS. They are:

1. Geometric image accuracy—the ability of the scanner to keep relative distances between points on an object (e.g., two minutiae) the same relative distances apart in the output image.
2. Modulation transfer function (MTF)—the ability of the scanning device to capture both low-frequency (ridges themselves) and high-frequency (ridge edge details) information in a fingerprint at minimum standards.
3. Signal-to-noise ratio—the ability of the scanning device to digitize the information without introducing too much electronic noise (that is, with the pure white image parts appearing pure white and the totally black image parts appearing totally black).
4. Gray-scale range of image data—avoiding excessively low-contrast images by ensuring that the image data are spread across a minimal number of shades of gray.
5. Gray-scale linearity—as the level of gray changes in a fingerprint capture, the digital image reflects a corresponding ratio of gray level across all shades of gray.
6. Output gray-level uniformity—the ability of the scanning device to create an image with a continuous gray scale across an area on the input image (tested using a special test image) that has a single gray level.

Interestingly, only two of these six image quality standards apply to latent scanning devices: geometric image accuracy and MTF. In fact, the FBI does not certify (see below for a discussion of certified products) scanners for latent use but recommends that latent examiners purchase equipment they are comfortable with using from an image-quality perspective. But EFTS Appendix F does mandate that latent images be captured at 1,000 ppi.

There are no standards for the quality of the actual fingerprint, but livescan and AFIS vendors have rated fingerprint quality for years. They know that fingerprint quality is possibly the strongest factor in the reliability of an AFIS's successfully matching a fingerprint to one in the repository. These ratings are often factored into the AFIS algorithms.

In a paper titled "The Role of Data Quality in Biometric Systems" (Hicklin and Khanna, 2006), the authors wrote the following:

> Note that this definition of data quality goes beyond most discussions of biometric quality, which focus on the concept of sample quality. Sample quality deals with the capture fidelity of the subject's physical characteristics and the intrinsic data content of those characteristics. However, an equally important issue for any operational system is metadata quality: databases need to be concerned with erroneous relationships between data elements, which generally come from administrative rather than biometric-specific causes.

Although no standard exists for fingerprint image quality, NIST has researched the relationship between calculated image quality (using algorithms similar to those employed by AFIS vendors) and successful match rates in automated fingerprint identification systems. This led NIST to develop and publish a software utility to measure fingerprint image quality.

The software is entitled NIST Fingerprint Image Software 2. It was developed by NIST's image group for the FBI and the U.S. Department of Homeland Security and is available free to U.S. law enforcement agencies as well as to biometrics manufacturers and researchers. The CD contains source code for 56 utilities and a user's guide.

The following summary is from the NIST Web site in 2007:

> New to this release is a tool that evaluates the quality of a fingerprint scan at the time it is made. Problems such as dry skin, the size of the fingers and the quality and condition of the equipment used can affect the quality of a print and its ability to be matched with other prints. The tool rates each scan on a scale from 1 for a high-quality print to 5 for an unusable one. "Although most commercial fingerprint systems already

include proprietary image quality software, the NIST software will for the first time allow users to directly compare fingerprint image quality from scanners made by different manufacturers," the agency said.

### *3.2.3. Certified Products List*

To assist the forensic community to purchase IQS-compliant equipment, the FBI established a certification program. The vendors can self-test their equipment and submit the results to the FBI where, with the technical assistance of Mitretek, the results are evaluated. If the results are acceptable, a letter of certification is sent to the vendor. It is important to know that, for capture devices, it is a combination of the optics (scanner), image processing software, and the operating system that is tested. Therefore, letters of certification are not issued for a scanner but for a scanner and PC configuration that includes a specific scanner model, connected to a PC running a specific operating system, and any image-enhancement scanner drivers used.

At the rate at which manufacturers upgrade scanners, it can be hard to purchase previously certified piccces of equipment. A complete list of all certified equipment is maintained on the FBI's Web site under the CJIS section.

### *3.2.4. Compression*

About the same time as the writing of the EFTS, the FBI decided on the compression standard for ANSI/NIST transmissions. Given that the data rate (bandwidth) of telecommunications systems was very low in 1993 compared to today's rates and that the cost of disk storage was quite high, the FBI elected to compress fingerprint images using a technique called wavelet scalar quantization (WSQ).

The initial plan was for tenprint transmissions to be compressed with WSQ at 20:1 and for latent images to remain uncompressed. An FBI fingerprint card in the early 1990s had a surface area for fingerprints that was 8 inches wide and 5 inches high for a total of 40 square inches.

Scanning at 500 ppi in both the 8-inch direction (X) and the 5-inch direction (Y) yielded a total of 10 million bytes of information (10 MB). Compression at 20:1 would produce a half (0.5) MB file that was much easier to transmit and store.

At the 1993 IAI Annual Training Conference in Orlando, Fl, the IAI Board of Directors expressed its concerns to the IAFIS program director about the proposed compression rate of 20:1. The FBI agreed to support an independent assessment of the impact of compression on the science of fingerprint identification by the IAI AFIS committee, under the Chairmanship of Mike

Fitzpatrick of Illinois (IAI AFIS Committee, 1994). As a result of the study, the FBI agreed to reduce the average compression to 15:1 (Higgins, 1995, pp 409–418).[5]

As other domains of interest adopted the ANSI/NIST standard around the world (early adopters included the Royal Canadian Mounted Police and the United Kingdom Home Office), they all used the EFTS as a model and all incorporated the IQS standard by reference. With one or two exceptions, they also adopted WSQ compression at 15:1.

With the move to higher scan rates for tenprint transactions, the compression technology of choice is JPEG 2000, which is a wavelet-based compression technique. Currently (as of 2007), there are at least five 1000-ppi tenprint, image-based automated fingerprint identification systems using JPEG 2000. Both Cogent and Motorola have delivered 1000-ppi systems. It is anticipated that the other vendors will deliver such systems as the demand increases. Given that older livescan systems operating at 500 ppi can submit transactions to these new automated fingerprint identification systems, it is important that they be capable of working in a mixed-density (500-ppi and 1000-ppi) environment.

All four major AFIS vendors demonstrated the capability to acquire, store, and process 1000-ppi tenprints and palmprints during the 2005 Royal Canadian Mounted Police AFIS Benchmark. It is important to note that these systems acquire the known tenprint and palm images at 1000 ppi for archiving but down-sample them to 500 ppi for searching and creating an image to be used in AFIS. Currently, 1000-ppi images are used primarily for display at latent examiner workstations. As automated fingerprint identification systems move to using third-level features, it is assumed that the higher resolution images will play a role in the algorithms.

### 3.3. Updates to the ANSI/NIST Standard

Since 1993, the ANSI/NIST standard has been updated three times, most recently in 2007 and 2008. The key changes are as follows:

- In 1997, type 10 transactions were added to permit facial, scar marks, and tattoo images to be transmitted with fingerprint transactions. The title of the document was changed to reflect that: "Data Format for the Interchange of Fingerprint, Facial & SMT (Scar, Mark, and Tattoo) Information (ANSI/NIST-ITL 1a-1997)".

- In 2000, types 13 through 16[6] were added to support higher density images, latent images in a new format, palm images, and test images, respectively (ANSI/ NIST-ITL 1-2000).
- NIST held two workshops in 2005 to determine whether there were any new areas that should be added. The major changes desired were the addition of standard record types for biometric data types beyond fingers and faces (e.g., iris images) and the introduction of XML data in the type 2 records. Several other changes and additions were also proposed. (See the 2007 and 2008 revisions, ANSI/NIST–ITL 1–2007 and 2–2008.)

## 3.4. Early Demonstrations of Interoperability

By 1996, the IAI AFIS Committee was organizing and managing (under the chairmanships of Mike Fitzpatrick, Peter Higgins, and Ken Moses) a series of demonstrations of interoperability of tenprint-image transactions originating from Aware software, Comnetix Live Scan, and Identix Live Scan and going to Cogent Systems, Printrak (now Motorola), and Sagem Morpho automated fingerprint identification systems. The second year of these demonstrations (1998) saw the same input being submitted between operational AFIS sites from the same three AFIS vendors all over the Nlets network (AFIS Committee Report, 1998, p 490).

## 3.5. Latent Interoperability

When IAFIS was being developed, the FBI established (in the EFTS) two ways for latent impressions to be run through IAFIS from outside agencies.

### *3.5.1. Remote Submittals*

The agency with the latent impression can send (electronically or via the mail) the impression (as an image in the case of electronic submittal) to the FBI, and FBI staff will perform the editing, encoding, searching, and candidate evaluation. The FBI will make any identification decision and return the results to the submitting agency. This process mimics the pre-IAFIS workflow but adds the option of electronic submittal.

#### *3.5.2. Remote Searches*

The agency with the latent impression performs the editing and encoding and then sends (electronically) a latent fingerprint features search (LFFS) to IAFIS for lights-out searching. IAFIS then returns a candidate list, including finger images, to the originating agency to perform candidate evaluation. The submitting agency makes any identification decision. To support LFFS remote search capability, the FBI published the "native" IAFIS feature set definition.

Many civil agencies and departments have wanted to be able to offer remote tenprint searches, but the feature sets for the major AFIS vendors are proprietary. In 2006, NIST performed a study on interoperability of the native feature set level of many AFIS and livescan companies and compared those with the performance of INCITS 378 fingerprint template standard minutiae (the basic set A and the richer set B).

The MINEX report (Grother et al., 2006) shows that minutiae-based interoperability is possible (with some loss of reliability and accuracy) for single-finger verification systems. The report is careful to point out that the use of INCITS 378 templates for remote criminal tenprint and latent searches is unknown and cannot safely be extrapolated from the report.

Because most AFISs (other than IAFIS) do not have remote LFFS functionality (as of 2007), latent interoperability at the image level usually requires labor on the part of the searching agency. The desire to move that labor burden to the submitting agency is natural because many have some level of excess capacity that could possibly support remote latent searches during off-hours.

# 4 Digitization and Processing of Fingerprints

## 4.1. Algorithms

Demands imposed by the painstaking attention needed to visually match the fingerprints of varied qualities, the tedium of the monotonous nature of the manual work, and increasing workloads due to a higher demand on fi ngerprint recognition services prompted law enforcement agencies to initiate research into acquiring fingerprints through electronic media and to automate fingerprint individualization based on digital representation of fingerprints. As a result of this research, a large number of computer algorithms have been developed during the past three decades to automatically process digital fingerprint images. An algorithm is a finite set of well-defined instructions for accomplishing some task which, given an initial state and input, will terminate in a corresponding recognizable end-state and output. A computer algorithm is an algorithm coded in a programming language to run on a computer. Depending upon the application, these computer algorithms could either assist human experts or perform in lights-out mode. These algorithms have greatly improved the operational productivity of law enforcement agencies and reduced the number of fingerprint technicians needed. Still, algorithm designers identified and investigated the following five major problems in designing automated fingerprint processing systems: digital fingerprint acquisition, image enhancement, feature (e.g., minutiae) extraction, matching, and indexing/retrieval.

## 4.2. Image Acquisition

Known fingerprint data can be collected by applying a thin coating of ink over a finger and rolling the finger from one end of the nail to the other end of the nail while pressing the finger against a paper card. This would result in an inked "rolled" fingerprint impression on the fingerprint card. If the finger was simply pressed straight down against the paper card instead of rolling, the resulting fi ngerprint impression would only contain a smaller central area of the finger rather than the full fingerprint, resulting in an inked "flat" or "plain" fingerprint impression.

The perspiration and contaminants on the skin result in the impression of a finger being deposited on a surface that is touched by that finger. These "latent" prints can be chemically or physically developed and electronically captured or manually "lifted" from the surface by employing certain chemical, physical, and lighting techniques. The developed fingerprint may be lifted with tape or photographed. Often these latent fingerprints contain only a portion of the friction ridge detail that is present on the finger, that is, a "partial" fingerprint.

Fingerprint impressions developed and preserved using any of the above methods can be digitized by scanning the inked card, lift, item, or photograph. Digital images acquired by this method are known as "off-line" images. (Typically, the scanners are not designed specifically for fingerprint applications.)

Since the early 1970s, fingerprint sensors have been built that can acquire a "livescan" digital fingerprint image directly from a finger without the intermediate use of ink and a paper card.

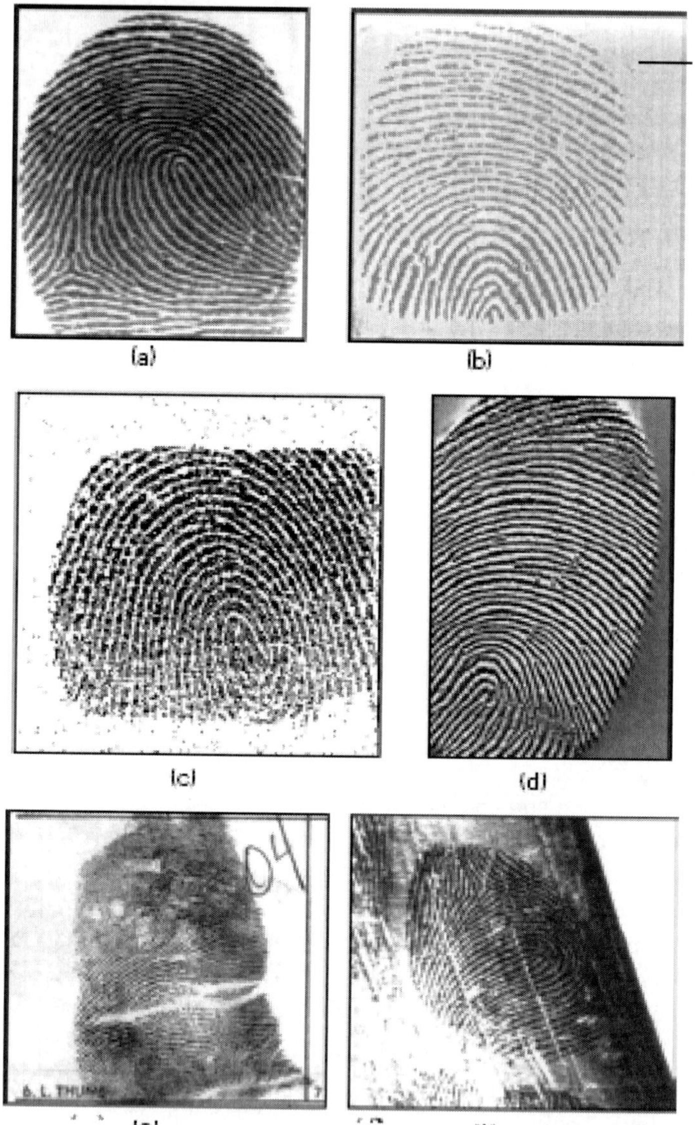


Figure 6. Fingerprint images from:

a) a livescan FTIR-based optical scanner; b) a livescan capacitive scanner; c) a livescan piezoelectric scanner; d) a livescan thermal scanner; e) an off-line inked impression; f) a latent fingerprint.

Although off-line images are still in use in certain forensic and government applications, on-line fingerprint images are increasingly being used. The main parameters characterizing a digital fingerprint image are resolution area, number of pixels, geometric accuracy, contrast, and geometric distortion. CJIS released specifications, known as Appendix F and Appendix G, that regulate the quality and the format of fingerprint images and FBI-compliant scanners. All livescan devices manufactured for use in forensic and government law enforcement applications are FBI compliant. Most of the livescan devices manufactured to be used in commercial applications, such as computer log-on, do not meet FBI specifications but, on the other hand, are usually more user-friendly, compact, and significantly less expensive. There are a number of livescan sensing mechanisms (e.g., optical, capacitive, thermal, pressure-based, ultrasound, and so forth) that can be used to detect the ridges and valleys present in the fingertip. However, many of these methods do not provide images that contain the same representation of detail necessary for some latent fingerprint comparisons. For example, a capacitive or thermal image may represent the edges and pores in a much different way than a rolled ink impression. Figure 6 shows an off-line fingerprint image acquired with the ink technique, a latent fingerprint image, and some livescan images acquired with different types of commercial livescan devices.

The livescan devices often capture a stream of fingerprint images from a single scan instead of just one image. Depending on the application for which the livescan device was designed, it may run one or more algorithms using either a resource-limited (memory and processing power) microprocessor on-board or by using an attached computer. For example, the livescan booking stations usually run an algorithm that can mosaic (stitch) multiple images acquired as a video during a single rolling of a finger on the scanner into a large rolled image. Algorithms also typically run on an integrated booking management system to provide real-time previews (graphical user interface and zoom) to assist the operator in placing or aligning fingers or palms correctly. Typically, a fingerprint image quality-checking algorithm is also run to alert the operator about the acquisition of a poor-quality fingerprint image so that a better quality image can be reacquired from the finger or palm. Typical output from such an automatic quality-checker algorithm is depicted in Figure 7.

Although optical scanners have the longest history and highest quality, the new solid-state sensors are gaining great popularity because of their compact size and the ease with which they can be embedded into laptop computers, cellular phones, smart pens, personal digital assistants (PDAs), and the like.

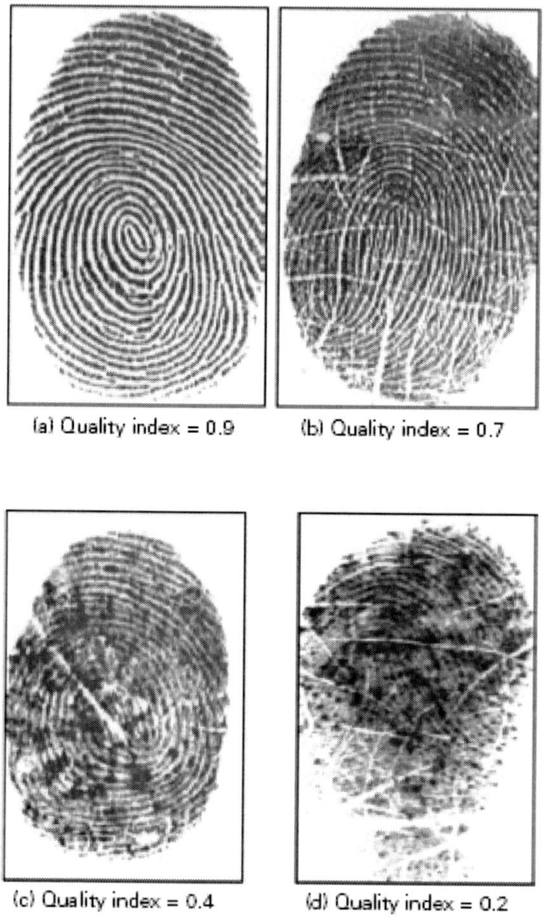


Figure 7. (a) A good-quality fingerprint; (b) A medium-quality fingerprint with creases; (c) A poor-quality fingerprint; (d) A very poor-quality fingerprint containing a lot of noise.

Swipe sensors, where a user is required to swipe his or her finger across a livescan sensor that is wide but very short, can offer the lowest cost and size. Such sensors image a single line or just a few lines (slice) of a fingerprint, and an image-stitching algorithm is used to stitch the lines or slices to form a two-dimensional fingerprint image (Figure 8).

Depending on the application, it may be desirable to implement one or more of the following algorithms in the livescan device:

- Automatic finger-detection algorithm—The scanner automatically keeps looking for the presence of a finger on its surface and, as soon as it determines that there is a finger present on its surface, it alerts the system.
- Automatic fingerprint-capture algorithm—Immediately after the system has been alerted that a finger is present on the surface of the scanner, it starts receiving a series of images, and the fingerprint-capture algorithm automatically determines which frame in the image sequence has the best image quality and grabs that frame from the video for further image processing and matching.
- Vitality detection algorithm—The scanner can determine whether the finger is consistent with deposition by a living human being.
- Image data-compression algorithm—Compressed image will require less storage and bandwidth when transferred to the system.
- Image-processing algorithms—Certain applications will benefit from feature extraction carried out on the sensor itself; the transfer of the fingerprint features will also require less bandwidth than the image.
- Fingerprint-matching algorithm—Certain applications would like the fingerprint matching to be performed on the sensor for security reasons, especially for on-board sequence checking.
- Cryptographic algorithms and protocol(s)—Implemented in the scanner to carry out secure communication.

## 4.3. Image Enhancement

Fingerprint images originating from different sources may have different noise characteristics and thus may require some enhancement algorithms based on the type of noise. For example, latent fingerprint images can contain a variety of artifacts and noise. Inked fingerprints can contain blobs or broken ridges that are due to an excessive or inadequate amount of ink. Filed paper cards may contain inscriptions overlapping the fingerprints and so forth. The goal of fingerprint enhancement algorithms is to produce an image that does not contain artificially generated ridge structure that might later result in the detection of false minutiae features while capturing the maximum available ridge structure to allow detection of true minutiae. Adapting the enhancement process to the fingerprint capture method can yield the optimal matching performance over a large collection of fingerprints.

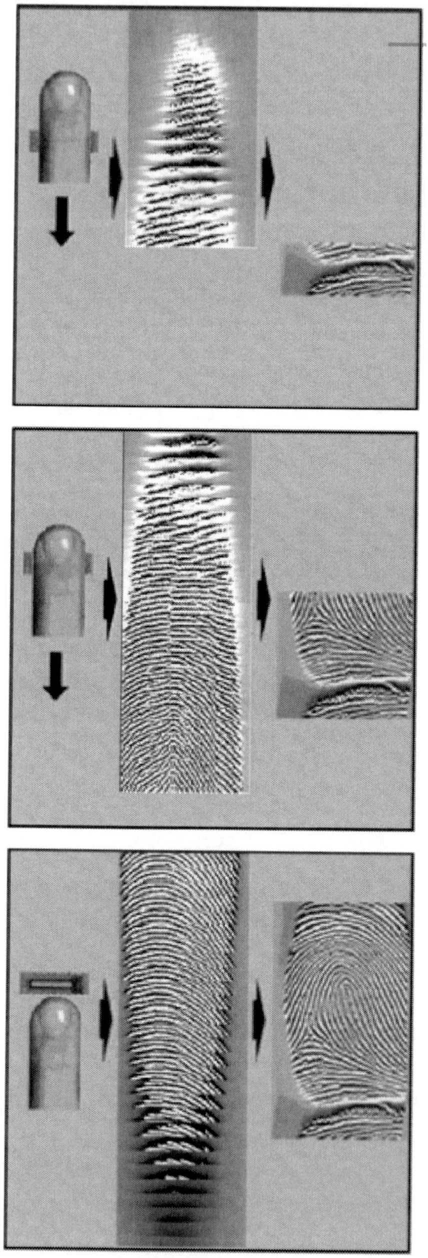

Figure 8. As the user sweeps his or her finger on the sensor, the sensor delivers new image slices, which are combined into a two-dimensional image.

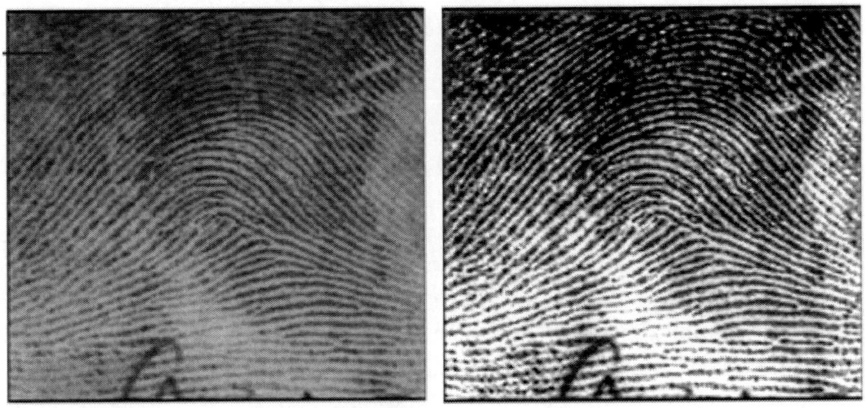

Figure 9. An example of local area contrast enhancement. The algorithm enhances the entire image by enhancing a large number of small square local areas.

A fingerprint may contain such poor-quality areas that the local ridge orientation and frequency estimation algorithms are completely wrong. An enhancement algorithm that can reliably locate (and mask) these extremely poor-quality areas is very useful for the later feature detection and individualization stages by preventing false or unreliable features from being created.

Fingerprint images can sometimes be of poor quality because of noise introduced during the acquisition process. For example: a finger may be dirty, a latent print may be lifted from a difficult surface, the acquisition medium (paper card or livescan) may be dirty, or noise may be introduced during the interaction of the finger with the sensing surface (such as slippage or other inconsistent contact). When presented with a poor-quality image, a forensic expert would use a magnifying glass and try to decipher the fingerprint features in the presence of the noise. Automatic fingerprint image-enhancement algorithms can significantly improve the quality of fingerprint ridges in the fingerprint image and make the image more suitable for further manual or automatic processing. The image enhancement algorithms do not add any external information to the fingerprint image. The enhancement algorithms use only the information that is already present in the fingerprint image. The enhancement algorithms can suppress various types of noise (e.g., another latent print, background color) in the fingerprint image and highlight the existing useful features. These image enhancement algorithms can be of two types.

#### *4.3.1. Enhancement of Latent Prints for AFIS Searching*

In the case of latent searches into the forensic AFISs, the enhancement algorithm is interactive, that is, live feedback about the enhancement is provided to the forensic expert through a graphical user interface. Through this interface, the forensic expert is able to use various algorithms to choose the region of interest in the fingerprint image, crop the image, invert color, adjust intensity, flip the image, magnify the image, resize the image window, and apply compression and decompression algorithms. The forensic expert can selectively apply many of the available enhancement algorithms (or select the parameters of the algorithm) based on the visual feedback. Such algorithms may include histogram equalization, image intensity rescaling, image intensity adjustments with high and low thresholds, local or global contrast enhancement, local or global background subtraction, sharpness adjustments (applying high-pass fi lter), background suppression (low-pass filter), gamma adjustments, brightness and contrast adjustments, and so forth. An example of local area contrast enhancement is shown in Figure 9. In this example, the fingerprint image enhancement algorithm enhances only a small, square, local area of the image at a time but traverses over the entire image in a raster scan fashion such that the entire image is enhanced. Subsequent fingerprint feature extraction can then be either performed manually or through automatic fingerprint feature extraction algorithms.

#### *4.3.2. Automated Enhancement of Fingerprint Images*

In the case of lights-out applications (frequently used in automated background checks and commercial applications for control of physical access), human assistance does not occur in the fingerprint individualization process. Enhancement algorithms are used in the fully automated mode to improve the fingerprint ridge structures in poor-quality fingerprint images.

An example of a fully automated fingerprint image- enhancement algorithm is shown in Figure 10. In this example, contextual filtering is used that has a low-pass (smoothing) effect along the fingerprint ridges and a band-pass (differentiating) effect in the direction orthogonal to the ridges to increase the contrast between ridges and valleys. Often, oriented band-pass filters are used for such filtering. One such type of commonly used filters is known as Gabor filters. The local context is provided to such contextual filters in terms of local orientation and local ridge frequency.

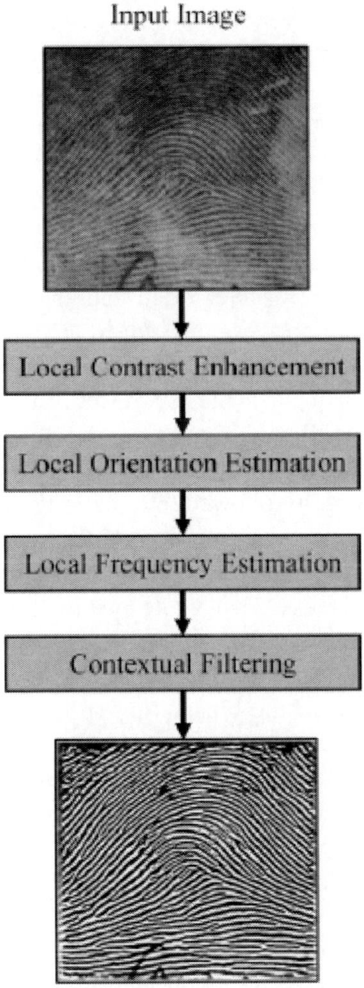


Figure 10. Stages in a typical contextual filtering-based fingerprint image enhancement algorithm.

## 4.4. Feature Extraction

Local fingerprint ridge singularities, commonly known as *minutiae points,* have been traditionally used by forensic experts as discriminating features in fingerprint images. The most common local singularities are ridge endings and ridge bifurcations. Other types of minutiae mentioned in the literature, such as the lake, island, spur, crossover, and so forth (with the exception of dots), are simply composites of ridge endings and bifurcations. Composite minutiae, made up of two to four minutiae occurring very close to each other, have also been used. In manual latent print processing, a forensic expert would visually locate the minutiae in a fingerprint image and note its location, the orientation of the ridge on which it resides, and the minutiae type. Automatic fingerprint feature-extraction algorithms were developed to imitate minutiae location performed by forensic experts. However, most automatic fingerprint minutiae-extraction algorithms only consider ridge endings and bifurcations because other types of ridge detail are very difficult to automatically extract. Further, most algorithms do not differentiate between ridge endings and bifurcations because they can be indistinguishable as a result of finger pressure differences during acquisition or artifacts introduced during the application of the enhancement algorithm.

One common approach followed by the fingerprint feature extraction algorithms is to first use a binarization algorithm to convert the gray-scale-enhanced fingerprint image into binary (black and white) form, where all black pixels correspond to ridges and all white pixels correspond to valleys. The binarization algorithm ranges from simple thresholding of the enhanced image to very sophisticated ridge location algorithms. Thereafter, a thinning algorithm is used to convert the binary fingerprint image into a single pixel width about the ridge centerline. The central idea of the thinning process is to perform successive (iterative) erosions of the outermost layers of a shape until a connected unit-width set of lines (or skeletons) is obtained. Several algorithms exist for thinning. Additional steps in the thinning algorithm are used to fill pores and eliminate noise that may result in the detection of false minutiae points.

The resulting image from the thinning algorithm is called a *thinned image* or *skeletal image.* A minutiae detection algorithm is applied to this skeletal image to locate the x and y coordinates as well as the orientation (theta) of the minutiae points. In the skeletal image, by definition, all pixels on a ridge have two neighboring pixels in the immediate neighborhood. If a pixel has only one neighboring pixel, it is determined to be a ridge ending and if a pixel has three neighboring pixels, it is determined to be a ridge bifurcation.

Each of the algorithms used in fingerprint image enhancement and minutiae extraction has its own limitation and results in imperfect processing, especially when the input fingerprint image includes non-friction-ridge noise. As a result, many false minutiae may be detected by the minutiae detection algorithm. To alleviate this problem, often a minutiae postprocessing algorithm is used to confirm or validate the detected minutiae. Only those minutiae that pass this postprocessing algorithm are kept and the rest are removed. For example, if a ridge length running away from the minutia point is sufficient or if the ridge direction at the point is within acceptable limits, the minutia is kept.

The postprocessing might also include an examination of the local image quality, neighboring detections, or other indicators of nonfingerprint structure in the area. Further, the image can be inverted in gray scale, converting white to black and black to white. Reprocessing of this inverted image should yield minutiae endings in place of bifurcations, and vice versa, allowing a validity check on the previously detected minutiae. The final detected minutiae are those that meet all of the validity checks. Figure 11 shows the steps in a typical fingerprint feature-extraction algorithm; the extracted minutiae are displayed overlapping on the input image for visualization.

Note that the stages and algorithms described in this section represent only a typical fingerprint minutiae-extraction algorithm. A wide variety of fingerprint minutiae-extraction algorithms exist and they all differ from one another, sometimes in how they implement a certain stage and sometimes in the stages they use and the order in which they use them. For example, some minutiae extraction algorithms do not use a postprocessing stage. Some others do not use a ridge-thinning stage, and the minutiae detection algorithm works directly on the result of the ridge location algorithm. Some work directly on the enhanced image, and some even work directly on the raw input image. Additional stages and algorithms may also be used.

Many other features are often also extracted in addition to minutiae. These additional features often provide useful information that can be used in the later matching stages to improve the fingerprint-matching accuracy. For example, minutiae confidence, ridge counts between minutiae, ridge count confidence, core and delta locations, local quality measures, and so forth, can be extracted. These additional features may be useful to achieve added selectivity from a minutiae-matching process. Their usefulness for this purpose may be mediated by the confidence associated with each such feature. Therefore, it is important to collect confidence data as a part of the image-enhancement and feature-extraction process to be able to properly qualify detected minutiae and associated features.

The early fingerprint feature-extraction algorithms were developed to imitate feature extraction by forensic experts. Recently, a number of automatic fingerprint feature- extraction (and matching) algorithms have emerged that use non-minutiae-based information in the fingerprint images.

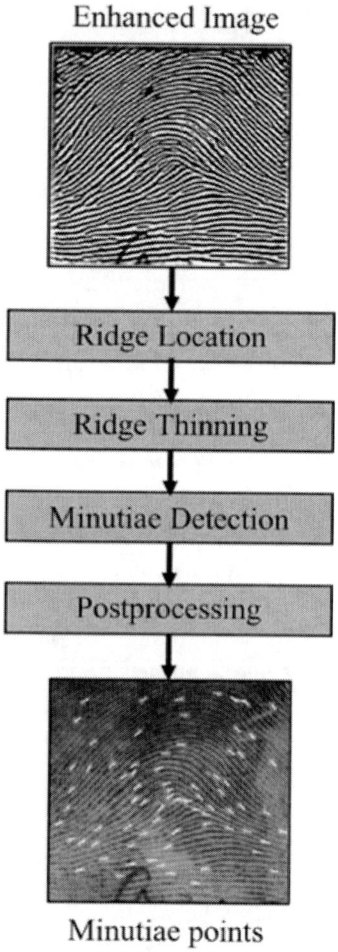


Figure 11. Stages in a typical fingerprint minutiae extraction algorithm.

For example, sweat pores, which are very minute details in fingerprints, smaller than minutiae points, have been successfully extracted by algorithms from high-resolution fingerprint images. Other non-minutiae-based features are often low-level features (for example, texture features) that do not have a high-level meaning, such as a ridge ending or bifurcation. These features are well suited for machine representation and matching and can be used in place of minutiae features. Often, a combination of minutiae and non-minutiae-based features can provide the best accuracy in an automatic fingerprint individualization system. Forensic experts use such fine features implicitly, along with normal ridge endings and bifurcations features, during examination.

## 4.5. Matching

Fingerprint matching can be defined as the exercise of finding the similarity or dissimilarity in any two given fingerprint images. Fingerprint matching can be best visualized by taking a paper copy of a file fingerprint image with its minutiae marked or overlaid and a transparency of a search fingerprint with its minutiae marked or overlaid. By placing the transparency of the search print over the paper copy of the file fingerprint and translating and rotating the transparency, one can locate the minutiae points that are common in both prints. From the number of common minutiae found, their closeness of fit, the quality of the fi ngerprint images, and any contradictory minutiae matching information, it is possible to assess the similarity of the two prints. Manual fingerprint matching is a very tedious task. Automatic fingerprint-matching algorithms work on the result of fingerprint feature-extraction algorithms and fi nd the similarity or dissimilarity in any two given sets of minutiae. Automatic fingerprint matching can perform fi ngerprint comparisons at the rate of tens of thousands of times each second, and the results can be sorted according to the degree of similarity and combined with any other criteria that may be available to further filter the candidates, all without human intervention.

It is important to note, however, that automatic fingerprint-matching algorithms are significantly less accurate than a well-trained forensic expert. Even so, depending on the application and the fingerprint image quality, the automaticfingerprint-matching algorithms can significantly reduce the work for forensic experts. For example, in the case of latent print matching where only a single, very poor quality partial fingerprint image is available for matching, the matching algorithm may not be very accurate. Still, the matching algorithm can return a list of candidate matches that is much smaller than the size of the database; the forensic expert then needs only to manually match a much smaller number of fingerprints. In the case of latent print matching when the latent print is of good quality, or in the case of tenprint-to-tenprint matching in a background check application, the matching is highly accurate and requires minimal human expert involvement.

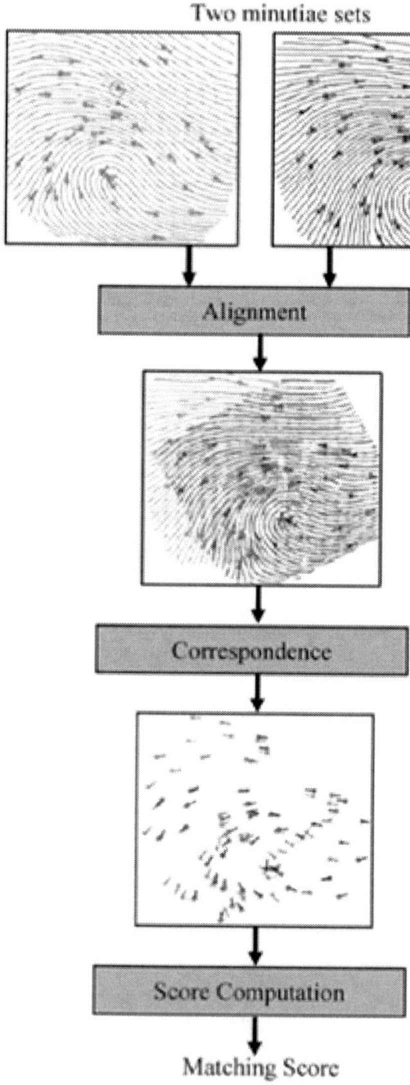


Figure 12. Stages in a typical fingerprint minutiae matching algorithm.

Automatic fingerprint-matching algorithms yield imperfect results because of the difficult problem posed by large intraclass variations (variability in different impressions of the same finger) present in the fingerprints. These intraclass variations arise from the following factors that vary during different acquisition of the same finger: (1) displacement, (2) rotation, (3) partial overlap, (4) nonlinear distortion because of pressing of the elastic three-dimensional finger onto a rigid two-dimensional imaging surface, (5) pressure, (6) skin conditions, (7) noise introduced by the imaging environment, and (8) errors introduced by the automatic feature-extraction algorithms.

A robust fingerprint-matching algorithm must be able to deal with all these intraclass variations in the various impressions of the same finger. The variations in displacement, rotation, and partial overlap are typically dealt with by using an alignment algorithm. The alignment algorithm should be able to correctly align the two fingerprint minutiae sets such that the corresponding or matching minutiae correspond well with each other after the alignment. Certain alignment algorithms also take into account the variability caused by nonlinear distortion. The alignment algorithm must also be able to take into consideration the fact that the feature extraction algorithm is imperfect and may have introduced false minutiae points and, at the same time, may have missed detecting some of the genuine minutiae points. Many fingerprint alignment algorithms exist. Some may use the core and delta points, if extracted, to align the fingerprints. Others use point pattern-matching algorithms such as Hough transform (a standard tool in pattern recognition that allows recognition of global patterns in the feature space by recognition of local patterns in a transformed parameter space), relaxation, algebraic and operational research solutions, "tree pruning," energy minimization, and so forth, to align minutiae points directly. Others use thinned ridge matching or orientation field matching to arrive at an alignment.

Once an alignment has been established, the minutiae from the two fingerprints often do not exactly overlay each other because of the small residual errors in the alignment algorithm and the nonlinear distortions. The next stage in a fingerprint minutiae-matching algorithm, which establishes the minutiae in the two sets that are corresponding and those that are noncorresponding, is based on using some tolerances in the minutiae locations and orientation to declare a correspondence. Because of noise that is introduced by skin condition, recording environment, imaging environment, and the imperfection of automatic fi ngerprint feature-extraction algorithms, the number of corresponding minutiae is usually found to be less than the total number of minutiae in either of the minutiae sets in the overlapping area. So, finally, a score computation algorithm is used to compute a matching score. The matching score essentially conveys the confidence of the fingerprint matching algorithm and can be viewed as an indication of the probability that the two fingerprints come from the same finger. The higher the matching score, the more likely it is that the fingerprints are mated (and, conversely, the lower the score, the less likely there is a match). There are many score computation algorithms that are used. They range from simple ones that count the number of matching minutiae normalized by the total number of minutiae in the two fingerprints in the overlapping area to very complex probability-theory-based, or statistical-pattern-recognitionclassifier-based algorithms that take into account a number of features such as the area of overlap, the quality of the fi ngerprints, residual distances between the matching minutiae, the quality of individual minutiae, and so forth. Figure 12 depicts the steps in a typical fingerprint matching algorithm.

Note that the stages and algorithms described in this section represent only a typical fingerprint minutiae-matching algorithm. Many fingerprint minutiae-matching algorithms exist and they all differ from one another. As with the various extraction algorithms, matching algorithms use different implementations, different stages, and different orders of stages. For example, some minutiae-matching algorithms do not use an alignment stage. These algorithms instead attempt to prealign the fingerprint minutiae so that alignment is not required during the matching stage. Other algorithms attempt to avoid both the prealignment and alignment during matching by defining an intrinsic coordinate system for fingerprint minutiae. Some minutiae-matching algorithms use local alignment, some use global alignment, and some use both local and global alignment. Finally, many new matching algorithms are totally different and are based on the non-minutiae-based features automatically extracted by the fingerprint feature-extraction algorithm, such as pores and texture features.

## 4.6. Indexing and Retrieval

In the previous section, the fingerprint matching problem was defined as finding the similarity in any two given fingerprints. There are many situations, such as controlling physical access within a location or affirming ownership of a legal document (such as a driver's license), where a single match between two fingerprints may suffice. However, in a large majority of forensic and government applications, such as latent fingerprint individualization and background checks, it is required that multiple fingerprints (in fact, up to 10 fingerprints from the 10 fingers of the same person) be matched against a large number of fingerprints present in a database. In these applications, a very large amount of fingerprint searching and matching is needed to be performed for a single individualization. This is very time-consuming, even for automatic fingerprint-matching algorithms. So it becomes desirable (although not necessary) to use automatic fingerprint indexing and retrieval algorithms to make the search faster.

Traditionally, such indexing and retrieval has been performed manually by forensic experts through indexing of fi ngerprint paper cards into file cabinets based on fingerprint pattern classification information as defined by a particular fingerprint classification system.

Similar to the development of the first automatic fingerprint feature extraction and matching algorithms, the initial automatic fingerprint indexing algorithms were developed to imitate forensic experts. These algorithms were built to classify fingerprint images into typically five classes (e.g., left loop, right loop, whorl, arch, and tented arch) based on the many fingerprint features automatically extracted from fingerprint images. (Many algorithms used only four classes because arch and tented arch types are often diffi cult to distinguish.)

Fingerprint pattern classification can be determined by explicitly characterizing regions of a fingerprint as belonging to a particular shape or through implementation of one of many possible generalized classifiers (e.g., neural networks) trained to recognize the specified patterns. The singular shapes (e.g., cores and deltas) in a fingerprint image are typically detected using algorithms based on the fingerprint orientation image. The explicit (rule-based) fi ngerprint classification systems first detect the fingerprint singularities (cores and deltas) and then apply a set of rules (e.g., arches and tented arches often have no cores; loops have one core and one delta; whorls have two cores and two deltas) to determine the pattern type of the fingerprint image (Figure 13). The most successful generalized (e.g., neural network-based) fingerprint classification systems use a combination of several different classifiers.

Such automatic fingerprint classification algorithms may be used to index all the fingerprints in the database into distinct bins (most implementations include overlapping or compared to only the database records with the same classification (i.e., in the same bin). The use of fingerprint pattern information can be an effective means to limit the volume of data sent to the matching engine, resulting in benefits in the system response time. However, the automatic fingerprint classification algorithms are not perfect and result in errors in classification. These classification errors increase the errors in fingerprint individualization because the matching effort will be conducted only in a wrong bin. Depending on the application, it may be feasible to manually confirm the automatically determined fingerprint class for some of the fingerprints where the automatic algorithm has low confidence. Even so, the explicit classification of fingerprints into just a few classes has its limitations because only a few classes are used (e.g., five), and the fingerprints occurring in nature are not equally distributed in these classes (e.g., arches and tented arches are much more rare than loops and whorls).

Many of the newer automatic fingerprint classification algorithms do not use explicit classes of fingerprints in distinct classifications but rather use a continuous classification of fingerprints that is not intuitive for manual processing but is amenable to automatic search algorithms. In continuous classification, fingerprints are associated with numerical vectors summarizing their main features. These feature vectors are created through a similarity-preserving transformation, so that similar fingerprints are mapped into close points (vectors) in the multidimensional space. The retrieval is performed by matching the input fingerprint with those in the database whose corresponding vectors are close to the searched one. Spatial data structures can be used for indexing very large databases. A continuous classification approach allows the problem of exclusive membership of ambiguous fingerprints to be avoided and the system's efficiency and accuracy to be balanced by adjusting the size of the neighborhood considered. Most of the continuous classification techniques proposed in the literature use the orientation image as an initial feature but differ in the transformation adopted to create the final vectors, and in the distance measure.

Some other continuous indexing methods are based on fingerprint minutiae features using techniques such as geometric hashing. Continuous indexing algorithms can also be built using other non-minutiae-based fingerprint features such as texture features.

Choosing an indexing technique alone is usually not sufficient; a retrieval strategy is also usually defined according to the application requirements, such as the desired accuracy and efficiency, the matching algorithm used to compare fingerprints, the involvement of a human reviewer, and so on. In general, different strategies may be defined for the same indexing mechanism. For instance, the search may be stopped when a fixed portion of the database has been explored or as soon as a matching fingerprint is found. (In latent fingerprint individualization, a forensic expert visually examines the fingerprints that are considered sufficiently similar by the minutiae matcher and terminates the search when a true correspondence is found.)

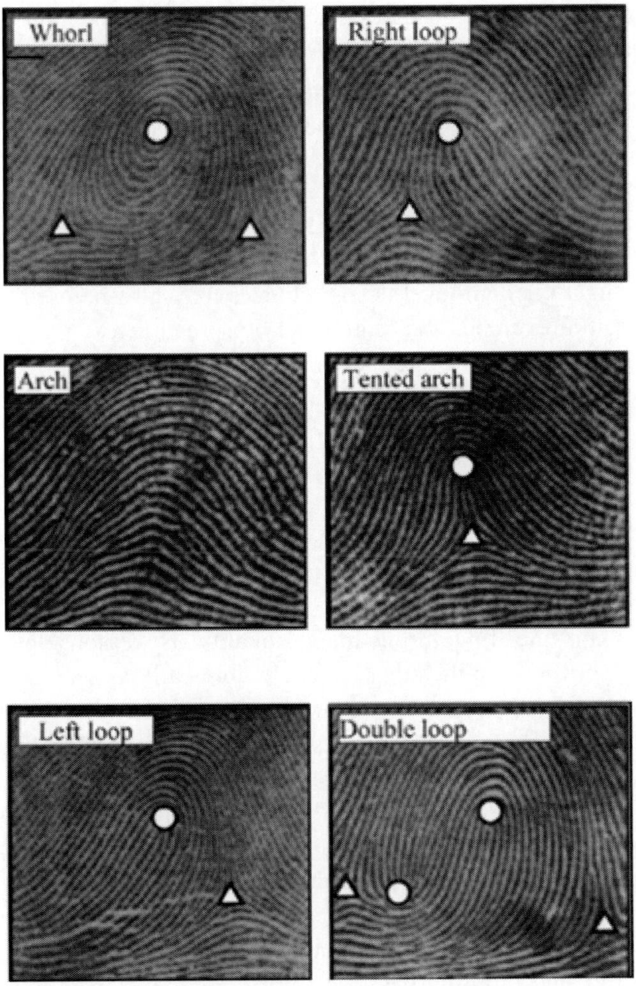


Figure 13. The six commonly used fingerprint classes: (a) whorl, (b) right loop, (c) arch, (d) tented arch, (e) left loop, and (f) double loop whorl.

If an exclusive classification technique is used for indexing, the following retrieval strategies can be used:

- Hypothesized class only—Only fingerprints belonging to the class to which the input fingerprint has been assigned are retrieved.
- Fixed search order—The search continues until a match is found or the whole database has been explored. If a correspondence is not

found within the hypothesized class, the search continues in another class, and so on.

- Variable search order—The different classes are visited according to the class likelihoods produced by the classifier for the input fingerprint. The search may be stopped as soon as a match is found or when the likelihood ratio between the current class and the next to be visited is less than a fixed threshold.

Finally, many system-level design choices may also be used to make the retrieval fast. For example, the search can be spread across many computers, and special- purpose hardware accelerators may be used to conduct fast fingerprint matching against a large database.

## 4.7. Accuracy Characterization

Although manual fingerprint matching is a very tedious task, a well-trained forensic expert is not likely to make individualization mistakes, especially when the fingerprint image quality is reasonable. Automatic fingerprint algorithms, on the other hand, are not nearly as accurate as forensic experts and have difficulty in dealing with the many noise sources in fingerprint images. Accuracy of fingerprint algorithms is crucial in designing fingerprint systems for real-world usage. The matching result must be reliable because many real-world decisions will be based on it. Algorithm designers usually acquire or collect their own fingerprint database and test the accuracy of their fingerprint algorithms on this database. By testing new algorithms, or changes in the old algorithm, or changes in algorithm parameters on the same database, they can know whether the new algorithm or changes improve the accuracy of the algorithm. Further, the algorithms' developers look closely at the false-positive and false-nonmatch errors made by their algorithms and get a better understanding of the strengths and limitations of their algorithms. By comparing the errors made by different algorithms or changes, the algorithm designers try to understand whether a change improves false positives, false nonmatches, both, or neither, and why. The algorithms' designers can then come up with algorithmic techniques to address the remaining errors and improve their algorithms' accuracy. It is desirable to have as large a database of fingerprints as possible from as large a demography as possible so that the algorithms are not overly adjusted to any certain variety of fingerprints and the accuracy obtained in the laboratory generalizes well in the field. Public organizations (e.g., NIST) perform periodic testing of fingerprint algorithms from different vendors on a common database to judge their relative accuracy.

There is a trade-off between the false positives and falsenonmatch error rates in fingerprint matching. Either of these two errors can be lowered at the expense of increasing the other error. Different applications have different requirements for these two types of errors. Interestingly, different fingerprint algorithms may perform differently, depending on the error rates. For example, algorithm A may be better than algorithm B at a low false-positive rate, but algorithm B may be better than algorithm A at a low false-nonmatch rate. In such cases, the algorithm designers may choose a certain algorithm or specific parameters to be used, depending on the application.

## 5. SUMMARY

Fingerprint technology has come a long way since its inception more than 100 years ago. The first primitive live- scan fingerprint readers introduced in 1988 were unwieldy beasts with many problems as compared to the sleek, inexpensive, and relatively miniscule sensors available today. During the past few decades, research and active use of fingerprint matching and indexing have also advanced our understanding of individuality, information in fingerprints, and efficient ways of processing this information. Increasingly inexpensive computing power, less expensive fingerprint sensors, and the demand for security, efficiency, and convenience have led to the viability of automatic fingerprint algorithms for everyday use in a large number of applications.

There are a number of challenges that remain to be overcome in designing a completely automatic and reliable fingerprint individualization system, especially when fingerprint images are of poor quality. Although automatic systems have improved significantly, the design of automated systems do not yet match the complex decision-making of a well-trained fingerprint expert as decisions are made to match individual fingerprints (especially latent prints). Still, automatic fingerprint matching systems hold real promise for the development of reliable, rapid, consistent, and cost-effective solutions in a number of traditional and newly emerging applications.

Research in automatic fingerprint recognition has been mostly an exercise in imitating the performance of a human fingerprint expert without access to the many underlying information-rich features an expert is able to glean by visual examination. The lack of such a rich set of informative features in automatic systems is mostly because of the unavailability of complex modeling and image-processing techniques that can reliably and consistently extract detailed features in the presence of noise. Perhaps using the human, intuition-based manual fingerprint recognition approach may not be the most appropriate basis for the design of automatic fingerprint recognition systems. There may be a need for exploring radically different features rich in discriminatory information, robust methods of fingerprint matching, and more ingenious methods for combining fingerprint matching and classification that are amenable to automation.

## 6. Reviewers

The reviewers critiquing this chapter were Patti Blume, Christophe Champod, Wayne Eaton, Robert J. Garrett, Laura A. Hutchins, Peter D. Komarinski, and Kasey Wertheim.

## Standard Terminology

***Preamble*** This document provides standard definitions for relevant terminology used in the friction ridge discipline. Common definitions found in other reference sources may not be included.

**ACE-V.** The acronym for a scientific method; Analysis, Comparison, Evaluation, and Verification (see individual terms).

**AFIS.** The acronym for Automated Fingerprint Identification System, a generic term for a fingerprint matching, storage, and retrieval system.

**Analysis.** The first step of the ACE-V method. The assessment of an impression to determine suitability for comparison.

**APIS.** The acronym for Automated Palmprint Identification System, a generic term for a palmprint (or complete friction ridge exemplar) matching, storage, and retrieval system.

**Arch – plain.** A pattern type in which the friction ridges enter on one side of the impression and flow, or tend to flow, out the other side with a rise or wave in the center.

**Arch - tented.** A pattern type that possesses either an angle, an upthrust, or two of the three basic characteristics of the loop.

**Artifact.** 1. Any distortion or alteration not in the original friction ridge impression, produced by an external agent or action.

2. Any information not present in the original object or image, inadvertently introduced by image capture, processing, compressions, transmission, display, or printing.

**Bias.** See cognitive bias, confirmation bias, and contextual bias.

**Bifurcation.** The point at which one friction ridge divides into two friction ridges.

**Blind verification.** The independent examination of one or more friction ridge impressions at any stage of the ACE process by another competent examiner who is provided with no, or limited, contextual information, and has no expectation or knowledge of the determinations or conclusions of the original examiner.

**Bridge.** A connecting friction ridge between, and generally at right angles to, parallel running friction ridges.

**Characteristics.** Distinctive details of the friction ridges, including Level 1, 2, and 3 details (also known as features).

**Cognitive bias.** The effect of perceptual or mental processes on the reliability and validity of one's observations and conclusions.

**Comparison.** The second step of the ACE-V method. The observation of two or more impressions to determine the existence of discrepancies, dissimilarities, or similarities.

**Competency.** Possessing and demonstrating the requisite knowledge, skills, and abilities to successfully perform a specific task.

**Complete friction ridge exemplars.** A systematic recording of all friction ridge detail appearing on the palmar sides of the hands. This includes the extreme sides of the palms, joints, tips, and sides of the fingers (also known as major case prints).

**Complex examinations.** The encountering of uncommon circumstances during an examination (e.g., the existence of high distortion, low quality or quantity, the possibility of simultaneity, or conflicts among examiners).

**Conclusion.** Determination made during the evaluation stage of ACE-V, including individualization, inconclusive, exclusion.

**Confirmation bias.** The tendency to search for data or interpret information in a manner that supports one's preconceptions.

**Conflict.** A difference of determinations or conclusions that becomes apparent during, or at the end of, an examination.

**Consultation.** A significant interaction between examiners regarding one or more impressions in question.

**Contextual bias.** The effect of information or outside influences on the evaluation and interpretation of data.

**Core.** 1. The approximate center of a fingerprint pattern.
2. A specific formation within a fingerprint pattern, defined by classification systems such as Henry.

**Delta.** The point on a friction ridge at or nearest to the point of divergence of two type lines, and located at or directly in front of the point of divergence. Also known as a tri-radius.

**Deviation.** 1. A change in friction ridge path.
2. An alteration or departure from a documented policy or standard procedure.

**Discrepancy.** The presence of friction ridge detail in one impression that does not exist in the corresponding area of another impression (compare with dissimilarity).

**Dissimilarity.** A difference in appearance between two friction ridge impressions (compare with discrepancy).

**Dissociated ridges.** 1. Disrupted, rather than continuous, friction ridges.
2. An area of friction ridge units that did not form into friction ridges, generally due to a genetic abnormality.

**Distortion.** Variances in the reproduction of friction skin caused by factors such as pressure, movement, force, and contact surface.

**Dot.** An isolated friction ridge unit whose length approximates its width in size.

**Edgeoscopy.** 1. Study of the morphological characteristics of friction ridges.
2. Contour or shape of the edges of friction ridges.

**Elimination prints.** Exemplars of friction ridge skin detail of persons known to have had legitimate access to an object or location.

**Enclosure.** A single friction ridge that bifurcates and rejoins after a short course and continues as a single friction ridge.

**Ending ridge.** A single friction ridge that terminates within the friction ridge structure.

**Erroneous exclusion.** The incorrect determination that two areas of friction ridge impressions did not originate from the same source.

**Erroneous individualization.** The incorrect determination that two areas of friction ridge impressions originated from the same source.

**Evaluation.** The third step of the ACE-V method wherein an examiner assesses the value of the details observed during the analysis and the comparison steps and reaches a conclusion.

**Exclusion.** The determination by an examiner that there is sufficient quality and quantity of detail in disagreement to conclude that two areas of friction ridge impressions did not originate from the same source.

**Exemplars.** The prints of an individual, associated with a known or claimed identity, and deliberately recorded electronically, by ink, or by another medium (also known as known prints).

**Features.** Distinctive details of the friction ridges, including Level 1, 2, and 3 details (also known as characteristics).

**Fingerprint.** An impression of the friction ridges of all or any part of the finger.

**Focal points.** 1. In classification, the core(s) and the delta(s) of a fingerprint.
2. Another term for target group.

**Friction ridge.** A raised portion of the epidermis on the pal-mar or plantar skin, consisting of one or more connected ridge units.

**Friction ridge detail (morphology).** An area comprised of the combination of ridge flow, ridge characteristics, and ridge structure.

**Friction ridge examiner.** A person who analyzes, compares, evaluates, and verifies friction ridge impressions.

**Friction ridge unit.** A single section of ridge containing one pore.

**Furrows.** Valleys or depressions between friction ridges.

**Galton details.** Term referring to friction ridge characteristics (also known as minutiae) attributed to the research of English fingerprint pioneer, Sir Francis Galton.

**Henry Classification.** An alpha-numeric system of fingerprint classification named after Sir Edward Richard Henry used for filing, searching, and retrieving tenprint records.

**IAFIS.** The acronym for Integrated Automated Fingerprint Identification System, the FBI's national AFIS.

**Identification.** 1. See individualization.
2. In some forensic disciplines, this term denotes the similarity of class characteristics.

**Impression.** Friction ridge detail deposited on a surface.

**Incipient ridge.** A friction ridge not fully developed that may appear shorter and thinner than fully developed friction ridges.

**Inconclusive.** The determination by an examiner that there is neither sufficient agreement to individualize, nor sufficient disagreement to exclude.

**Individualization.** The determination by an examiner that there is sufficient quality and quantity of detail in agreement to conclude that two friction ridge impressions originated from the same source.

**Joint (of the finger).** The hinged area that separates segments of the finger.

Page 7 of 11

**Known prints (finger, palm, foot).** The prints of an individual, associated with a known or claimed identity, and deliberately recorded electronically, by ink, or by another medium (also known as exemplars).

**Latent print.** 1. Transferred impression of friction ridge detail not readily visible.

2. Generic term used for unintentionally deposited friction ridge detail.

**Level 1 detail.** Friction ridge flow, pattern type, and general morphological information.

**Level 2 detail.** Individual friction ridge paths and associated events, including minutiae.

**Level 3 detail.** Friction ridge dimensional attributes, such as width, edge shapes, and pores.

**Lift.** An adhesive or other medium used to transfer a friction ridge impression from a substrate.

**Loop.** A pattern type in which one or more friction ridges enter upon one side, recurve, touch or pass an imaginary line between delta and core and flow out, or tend to flow out, on the same side the friction ridges entered. Types include left slant loops, in which the pattern flows to the left in the impression; right slant loops, in which the pattern flows to the right in the impression; radial loops, in which the pattern flows in the direction of the radius bone of the forearm (toward the thumb); and ulnar loops, in which the pattern flows in the direction of the ulna bone of the forearm (toward the little finger).

**Major case print.** A systematic recording of the friction ridge detail appearing on the palmar sides of the hands. This includes the extreme sides of the palms, joints, tips, and sides of the fingers (also known as complete friction ridge exemplars).

**Mark.** Term commonly used in the United Kingdom and some Commonwealth countries to designate a latent print.

**Matrix.** The substance that is deposited or removed by the friction ridge skin when making an impression.

**Minutiae.** Events along a ridge path, including bifurcations, ending ridges, and dots (also known as Galton details).

**Missed individualization.** The failure to make an individualization when in fact both friction ridge impressions are from the same source.

**NGI.** The acronym for Next Generation Identification, the updated version of IAFIS.

**Original image.** Primary image; with respect to digital images, an accurate replica (bit-for-bit value) of the primary image.

**Palmprint.** An impression of the friction ridges of all or any part of the palmar surface of the hand.

**Pattern classification.** Sub-division of pattern type, defined by classification systems such as Henry or National Crime Information Center (NCIC) classifications.

**Pattern type**. Fundamental pattern of the ridge flow: arch, loop, whorl. Arches are subdivided into plain and tented arches; loops are subdivided into radial and ulnar loops; whorls are subdivided into plain whorls, double loops, pocket loops, and accidental whorls.

**Phalanx/Phalange.** 1. A bone of the finger or toe.
2. Sometimes used to refer to a segment of a finger.

**Poroscopy.** A study of the size, shape, and arrangement of pores.

**Primary image.** The first recording of an image onto media.

**Proficiency.** The ongoing demonstration of competency.

**Quality.** The clarity of information contained within a friction ridge impression.

**Quantity.** The amount of information contained within a friction ridge impression.

**Ridge flow.** 1. The direction of one or more friction ridges. 2. A component of Level 1 detail.

**Ridge path.** 1. The course of a single friction ridge. 2. A component of Level 2 detail. **Ridge unit.** See friction ridge unit.

**Segment (of the finger).** The proximal, medial, or distal section of the finger.

**Short ridge.** A single friction ridge beginning, traveling a short distance, and then ending.

**Simultaneous impression**. Two or more friction ridge impressions from the same hand or foot deposited concurrently.

**Source.** An area of friction ridge skin from an individual from which an impression originated.

**Spur.** A bifurcation with one short friction ridge branching off a longer friction ridge.

**Stand-alone.** A segment of a simultaneous impression that has sufficient information to arrive at a conclusion of individualization independent of other impressions within the aggregate.

**Substrate.** The surface upon which a friction ridge impression is deposited.

**Sufficiency.** The product of the quality and quantity of the objective data under observation (e.g., friction ridge, crease, and scar features).

**Sufficient.** The determination that there is sufficiency in a comparison to reach a conclusion at the evaluation stage.

**Suitable.** The determination that there is sufficiency in an impression to be of value for further analysis or comparison.

**Target group.** A distinctive group of ridge features (and their relationships) that can be recognized.

**Technical review.** Review of notes, documents, and other data that forms the basis for a scientific conclusion (see ASCLDLAB 2008 Manual).

**Tenprint.** 1. A generic reference to examinations performed on intentionally recorded friction ridge impressions.

2. A controlled recording of an individual's available fingers using ink, electronic imaging, or other medium.

**Tolerance.** The amount of variation in appearance of friction ridge features to be allowed during a comparison, should a corresponding print be made available.

**Trifurcation.** The point at which one friction ridge divides into three friction ridges.

**Type lines.** The two innermost friction ridges associated with a delta that parallel, diverge, and surround or tend to surround the pattern area.

**Verification.** The independent application of the ACE process as utilized by a subsequent examiner to either support or refute the conclusions of the original examiner; this may be conducted as blind verification. Verification may be followed by some level of review as specified by agency policy.

**Whorl - accidental.** 1. A pattern type consisting of the combination of two different types of patterns (excluding the plain arch) with two or more deltas.

2. A pattern type that possesses some of the requirements for two or more different types of patterns.

3. A pattern type that conforms to none of the definitions of a pattern.

**Whorl - central pocket loop.** A pattern type that has two deltas and at least one friction ridge that makes, or tends to make, one complete circuit, which may be spiral, oval, circular, or any variant of a circle. An imaginary line drawn between the two deltas must not touch or cross any re-curving friction ridges within the inner pattern area.

**Whorl - double loop.** A pattern type that consists of two separate loop formations with two separate and distinct sets of shoulders and two deltas.

**Whorl - plain.** A fingerprint pattern type that consists of one or more friction ridges that make, or tends to make, a complete circuit, with two deltas, between which, when an imaginary line is drawn, at least one recurving friction ridge within the inner pattern area is cut or touched.

## References

*AFIS Committee Report.* International Association for Identification: Mendota Heights, MN, 1994.

AFIS Committee Report. *J. Forensic Ident.* 1998, *48* (4), 489–500.

*American National Standards for Information Systems— Data Format for the Interchange of Fingerprint Information;* ANSI/NIST-CSL 1-1993; National Institute of Standards and Technology, U.S. Government Printing Office: Washington, DC, 1993.

*American National Standard for Information Systems— Data Format for the Interchange of Fingerprint, Facial & SMT (Scar, Mark, and Tattoo) Information;* ANSI/NIST-ITL 1a-1997; National Institute of Standards and Technology, U.S. Government Printing Office: Washington, DC, 1997.

*American National Standard for Information Systems— Data Format for the Interchange of Fingerprint, Facial, & & Scar Mark & Tattoo (SMT);* ANSI/NIST-ITL 1-2000, NIST Special Publication #500-245; National Institute of Standards and Technology, U.S. Government Printing Office: Washington, DC, 2000.

*American National Standards for Information Systems— Data Format for the Interchange of Fingerprint, Facial, & Other Biometric Information—Part 1 (Traditional Format);* ANSI/NIST-ITL 1-2007, NIST Special Publication #500-271; National Institute of Standards and Technology, U.S. Government Printing Office: Washington, DC, 2007. Available online at http://www.nist.gov/customcf/get_pdf.cfm?pub_ id=51174.

*American National Standards for Information Systems— Data Format for the Interchange of Fingerprint, Facial, & Other Biometric Information—Part 2 (XML Version);* ANSI/ NIST-ITL 2-2008, NIST Special Publication #500-275; National Institute of Standards and Technology, U.S. Government Printing Office: Washington, DC, 2008. Available online at http://www.nist.gov/customcf/get_pdf.cfm?pub_ id=890062.

Bruton, T. *Annual Report of the Crime Scene Investigations Unit;* San Francisco Police Department: San Francisco, CA, 1989.

Cole, S. *Suspect Identities;* Harvard University Press: Cambridge, MA, 2001.

FBI Request for Quotation No. 66-1, December 16, 1966.

Greenwood, P. W.; Chaiken, J. M.; Petersilia, J. *The Criminal Investigative Process* (Vols. 1–3); Technical Report R-1777-DOJ; RAND Corporation: Santa Monica, CA, 1975.

Grother, P.; McCabe, M.; et al. *MINEX: Performance and Interoperability of INCITS 378 Fingerprint Template;* NISTIR 7296; National Institute of Standards and Technology, March 21, 2006.

Hicklin, A.; Khanna, R. *The Role of Data Quality in Biometric Systems;* Mitretek Systems: Falls Church, VA, 2006.

Higgins, P. Standards for the Electronic Submission of Fingerprint Cards to the FBI. *J. Forensic Ident.* 1995, *45* (4), 409–418.

International Association for Identification. *AFIS Directory of Users;* IAI: Mendota Heights, MN, 1999.

Kiji, K. *AFIS 30-Year History;* NEC Internal Corporate Report; NEC Solutions: Tokyo, Japan, 2002.

Komarinski, P. *Automated Fingerprint Identification Systems;* Elsevier: New York, 2005.

Moore, R. T. Automatic Fingerprint Identification Systems. In *Advances in Fingerprint Technology,* 1st ed.; Lee, H. C.; Gaensslen, R. E., Eds.; Elsevier, NY, 1991; pp 163–191.

Moses, K. R. Consumer's Guide to Fingerprint Systems. *Ident. News* 1986, *36* (6), 5–7, 10.

National Institute for Standards and Technology. *MINEX: Performance and Interoperability of INCITS 378 Fingerprint Template (NISTIR 7296);* March 6, 2005.

Petersilia, J. *The Collection and Processing of Physical Evidence;* WN-9062-DOJ; RAND Corporation: Santa Monica, CA, 1975.

Wayman, J. *Biometric Systems.* Springer: New York, 2004.

Wegstein, J. H. *A Computer Oriented Single-Fingerprint Identification System;* Technical Note 443; National Bureau of Standards, U.S. Department of Commerce: Washington, DC, 1969a.

Wegstein, J. H. *A Semi-Automated Single Fingerprint Identification System;* Technical Note 481; National Bureau of Standards, U.S. Department of Commerce: Washington, DC, 1969b.

Wegstein, J. H. *Automated Fingerprint Identification;* Technical Note 538; National Bureau of Standards, U.S. Department of Commerce: Washington, DC, 1970.

Wegstein, J. H. *The M40 Fingerprint Matcher;* Technical Note 878; National Bureau of Standards, U.S. Department of Commerce: Washington, DC, 1972a.

Wegstein, J. H. *Manual and Automated Fingerprint Registration;* NBS Technical Note 730; National Bureau of Standards, U.S. Department of Commerce: Washington, DC, 1972b.

Wegstein, J. H. *An Automated Fingerprint Identification System;* NBS Special Publication 500-89; National Bureau of Standards, U.S. Department of Commerce: Washington, DC, 1982.

Wegstein, J. H.; Rafferty, J. F. *The LX39 Latent Fingerprint Matcher;* Special Publication 500-36; National Bureau of Standards, U.S. Department of Commerce: Washington, DC, 1978.

Wegstein, J. H.; Rafferty, J. F. The Automated Identification of Fingerprints. In *Dermatoglyphics—Fifty Years Later;* March of Dimes: Washington, DC, 1979.

Wegstein, J. H.; Rafferty, J. F.; Pencak, W. J. *Matching Fingerprints by Computer;* Technical Note 466; National Bureau of Standards, U.S. Department of Commerce: Washington, DC, 1968.

## Additional Information

Asai, K.; Kato, Y.; Hoshino, Y.; Kiji, K. Automatic Fingerprint Identification; In Proceedings of the SPIE, vol. 182—Imaging Applications for Automated Industrial Inspection and Assembly, 1979; pp 49–56.

Lee, H. C.; Gaensslen, R.E., Eds. Advances in Fingerprint Technology; 2nd ed.; CRC Press: Washington, D.C., 2001.

Millard, K. An Approach to the Automatic Retrieval of Latent Fingerprints; In Proceedings of Carnahan Conference on Electronic Crime Countermeasures, Lexington, KY, 1975; pp 45–51.

Millard, K. Development on Automatic Fingerprint Recognition; In Proceedings of the Carnahan Conference on Security Technology, Zurich, Switzerland, 1983; pp 173–178.

Prabhakar, S.; Jain, A.; Maltoni, D.; Maio, D. Handbook of Fingerprint Recognition; Springer-Verlag: New York, 2003.

Ratha, N.; Bolle, R., Eds. Automated Fingerprint Recognition Systems; Springer-Verlag: New York, 2004.

Roberts, D. F. Dermatoglyphics and Human Genetics. In Dermatoglyphics—Fifty Years Later; Birth Defects Original Article Series; Wertelecki, W., Plato, C., Paul, N. W., Eds.; Alan R. Liss Inc.: New York, 1979; pp 475–494.

Thiebault, R. Automatic Process for Automated Fingerprint Identification; In Proceedings of the International Symposium on Automation of Population Register Systems, 1967; pp 207–226.

Thiebault, R. An Automatic Procedure for Identifying Fingerprints. *International Criminal Police Rev.* 1970, 25, 2–10.

Uchida, K. Fingerprint Identification. *NEC J. Advanced Technology* 2005, 2 (1), 19–27.

Wayman, J.; Jain, A.; Maltoni, D.; Maio, D., Eds. Biometric Systems, Springer-Verlag: New York, 2005.

## End Notes

[1] NIST was known as the National Bureau of Standards when the FBI visited Moore and Wegstein.

[2] The results of the earliest competitive benchmark tests were published by the International Association for Identification in 1986 (Moses, 1986). Thereafter, some vendors often demanded that the results of benchmark tests be kept secret, and law enforcement agencies generally acquiesced to those demands. This has made it extremely difficult for researchers and prospective purchasers to evaluate competing systems. The veil of secrecy has generally carried over to the sharing of AFIS operational performance data by agency personnel who often develop a strong sense of loyalty to their AFIS vendor.

[3] This figure is based on requirements found in award documents and benchmark testing rather than operational observation.

[4] "Lights out" generally refers to the ability of the system to operate without human intervention.

[5] The study showed that expert latent print examiners were unable to differentiate original images from those compressed at either 5:1 or 10:1 when presented with enlargements on high-quality film printers. One possible implication of that study was that latent images might safely be compressed at 2:1 (or possibly even more) for transmission, with no loss of information content. Currently, there are no agencies reporting the use of compression with latent images.

[6] Types 11 and 12 were set aside for a project that never came to fruition and are not used in the standard AFIS Committee Report, 1998.

In: Automated Fingerprint Identification
Editor: Edward P. Jackson
ISBN: 978-1-63483-304-2


*Chapter 2*

# ACHIEVING INTEROPERABILITY FOR LATENT FINGERPRINT IDENTIFICATION IN THE UNITED STATES*

***Subcommittee on Forensic Science***

## ABBREVIATIONS

| | |
|---|---|
| ABIS | Automated Biometric Identification System |
| AFIS | Automated Fingerprint Identification System |
| ANSI | American National Standards Institute |
| BIMA | Biometrics Identity Management Activity |
| BJA | Bureau of Justice Assistance |
| CBP | Customs and Border Protection |
| CJIS | Criminal Justice Information Services |
| DHS | Department of Homeland Security |
| DLC | Direct Latent Connectivity |
| DOD | Department of Defense |
| EBTS | Electronic Biometric Transmission Standard |
| EFS | Extended Feature Set |
| FBI | Federal Bureau of Investigation |

* This is an edited, reformatted and augmented version of a document issued by the National Science and Technology Council (within the Office of Science and Technology Policy), April 2015.

| | |
|---|---|
| IAFIS | Integrated Automated Fingerprint Identification System |
| IDENT | Automated Biometric Identification System |
| INS | Immigration and Naturalization Service |
| ITL | Information Technology Laboratory |
| LEEP | Law Enforcement Enterprise Portal |
| *LFIOS* | *Latent Fingerprint Interoperability Survey* |
| LITS | Latent Interoperability Transmission Specification |
| ME/C | medical examiners and coroners |
| MOU | memorandum of understanding |
| NDR | Name of Designated Repository |
| NGI | Next Generation Identification |
| NIJ | National Institute of Justice |
| NIST | National Institute of Standards and Technology |
| Nlets | National Law Enforcement Telecommunications Systems |
| NOVARIS | Northern Virginia Regional Identification System |
| OSTP | Office of Science and Technology Network |
| RFP | Request for Proposals |
| RISC | Repository of Individuals of Special Concern |
| SIB | State Identification Bureau |
| SOP | Standard Operating Procedure |
| ULF | Unsolved Latent File |
| ULW | Universal Latent Workstation |
| US-VISIT | United States Visitor and Immigrant Status Indicator Technology |
| VPN | Virtual Private Network |
| TXDPS | Texas Department of Public Safety |
| WAN | Wide Area Network |
| WIN | Western Identification Network |

## EXECUTIVE SUMMARY

Automated Fingerprint Identification System (AFIS) interoperability will support public safety throughout the United States by ensuring local law enforcement agencies are better able to coordinate their investigative fingerprinting efforts. The Federal Government is facilitating AFIS interoperability by implementing standards, developing an overarching national connectivity strategy and infrastructure, and supporting state and local agencies in building connections across jurisdictions. This report describes the current state of latent AFIS interoperability and identifies actions that can be taken by Federal agencies to support the following:

- Acquisition of standards-compliant systems at the Federal, state, and local-levels;
- Furthering connectivity efforts among law enforcement agencies;
- Improved governance structures to reflect the new interoperable environment;
- Developing mechanisms to test system performance and standards compliance; and
- Expanded examiner training.

For over a century, fingerprints have been used among other applications to identify criminal and terrorist suspects, perform background checks, and monitor immigration status. In the context of a criminal investigation, prints found at the scene of a crime, known as "latent prints," are compared with fingerprint records of known individuals who have been convicted or arrested for a crime. A latent print is an impression of the friction ridges of any part of a human finger. Latent prints include fingerprints that are not readily visible and are recovered from evidence or a crime scene. The highly variable characteristics of latent prints complicate the identification process.

When law enforcement agencies first began using fingerprints in investigations, the comparison process was entirely manual and based on visual inspections of features found on both the latent print and the fingerprint records. AFIS, first introduced in the 1970s, uses image processing algorithms to generate a list of potential candidates that share fingerprint features similar to an encoded image of a latent print. The algorithms assess "minutiae" and other friction ridge features found on the underside of the finger and on the palm.

Developers of AFIS software differentiate themselves from their market competitors by independently creating algorithms to extract features and match images of varying latent print quality. These proprietary approaches have resulted in insufficient interoperability among different AFIS systems to meet the needs of law enforcement agencies.

"Interoperability" is the ability of two or more networks, systems, devices, applications, or components to use standardized encoding to seamlessly and electronically share information on demand and as authorized, without special effort and without loss of accuracy. Interoperability of AFIS software would allow law enforcement agencies to search fingerprint records beyond those within their own jurisdictions in support of efforts to identify suspects and protect public safety.

True interoperability requires technical compatibility, network connectivity, proper governance, and performance testing and training within and between systems. A national effort to improve interoperability in pursuit of public safety objectives is underway in each of these areas.

*Technical compatibility.* Technical compatibility is necessary for AFIS systems to communicate electronic data. It has been mostly achieved for ten-print searches and is advancing for latent fingerprint searches through the development of standards. Two recent standards that have advanced technical compatibility are the Extended Feature Set (EFS), which defines a common feature format, and the Latent Interoperability Transmission Specification (LITS), which delineates what information is needed in a transaction between systems.[1] The Federal Bureau of Investigation (FBI) recently introduced the Next Generation Identification (NGI) system, which has incorporated the EFS as the standard submission format for all of its latent print searches. The success of these standards now depends on their adoption by state and local agencies, which is presently underway.

*Network connectivity.* Law enforcement agencies' coordinated efforts to identify suspects across jurisdictional lines depend on the connectivity of different AFIS systems through established networks. At present, latent print data sharing is limited. State agencies have an AFIS that has direct access to the NGI system via the Criminal Justice Information Services (CJIS) Wide Area Network (WAN), a network capable of transmitting latent searches to and from the NGI system. Efforts are being explored to connect state agencies to one another and to connect local agencies to the NGI system through their State Identification Bureau (SIB) for latent searches.

*Proper governance.* Proper governance is required for interoperable sharing agreements so that agencies are transparent about their search activities, respect relevant privacy laws, and outline the conditions of cross-jurisdictional data sharing. Interagency coordination has been facilitated by guidelines on the process for developing agreements between agencies and the recommended language for these agreements, but many jurisdictions have not yet pursued agreements.

*Performance testing and training.* Underpinning the effectiveness of an interoperable system is assurance that the examiners are adequately trained and that the software complies with newly adopted standards, provides accurate results, and functions properly. Quality assurance of examiners has improved through the introduction of an online EFS training tool that allows examiners and AFIS vendors to explore the functionality of the EFS.

*Education and outreach.* Education and outreach are important overarching components for bringing about interoperability. Expanding outreach will support other Federal efforts to ensure AFIS interoperability and speed the adoption of standards.

## INTRODUCTION

This report, *Achieving Interoperability for Latent Fingerprint Identification in the United States,* provides the current landscape of latent Automated Fingerprint Identification System (AFIS) interoperability, outlines recent advances in the field, and provides a high-level strategic plan for Federal agencies to implement the necessary requirements for interoperability to enhance public safety. The report addresses five essential elements of interoperability—technical compatibility, network connectivity, proper governance, performance testing and training, and education and outreach. It also examines existing mechanisms for sharing information and coordinating fingerprint searches and outlines a path for building on best practices to achieve a national latent print interoperability infrastructure.

For over a century, fingerprints have been used to identify criminal and terrorist suspects, to perform background checks, to monitor immigration status, to support public safety, and for other purposes. In the context of a criminal investigation, prints found at the scene of a crime, referred to as latent prints, are compared with fingerprint records of known individuals who have been convicted of or arrested for a crime. When law enforcement agencies first began using fingerprints in investigations, the comparison process was entirely manual and based on visual inspections of features found on both the latent print and the fingerprint records.

Law enforcement agencies began working with industry to automate this comparison process in the late 1970s and early 1980s by employing AFIS systems. After a latent print search has been initiated, AFIS systems use image-recognition algorithms to generate a list of potential candidates who share similar fingerprint features to an encoded image of a latent fingerprint. The algorithms assess "minutiae" and other friction ridge features found on the underside of the finger and on the palm. To submit a search, fingerprint examiners manually encode ("markup") an image to indicate the locations of minutiae on a digital image of the print. Law enforcement agencies have benefitted from the adoption of AFIS systems because potential suspects can be identified more quickly and accurately, often when no methods.

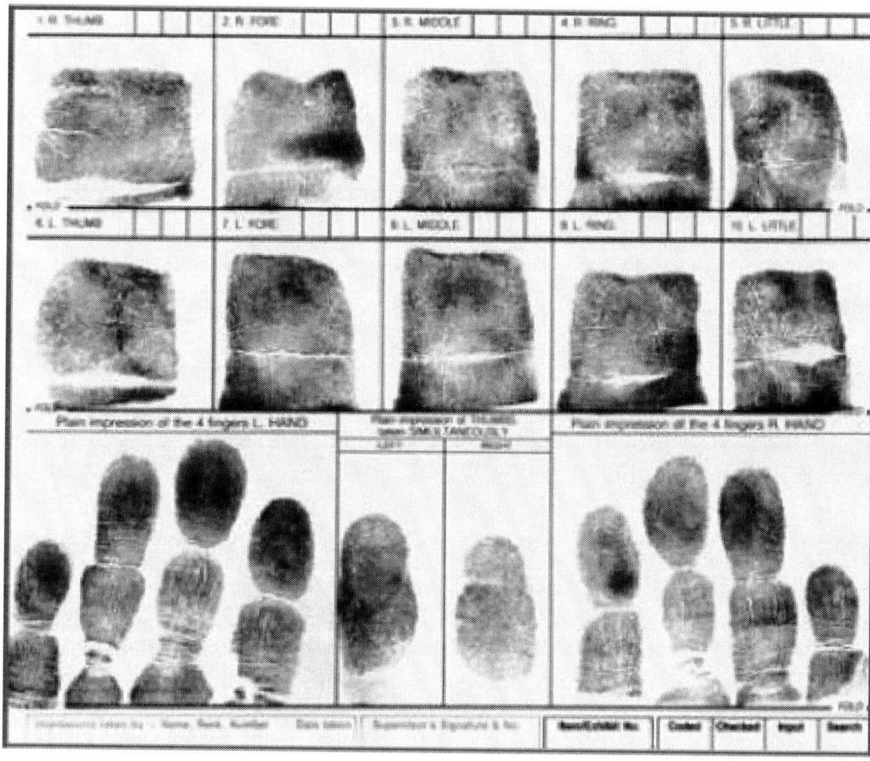

Figure 1. Example of a ten-print record.

Law enforcement agencies perform two types of fingerprint identification using AFIS systems: exemplar searches (more commonly known as "ten-print" searches) and latent print searches. A ten-print record contains fingerprint images of up to ten fingers captured by law enforcement agencies or private entities in a controlled environment as a result of arrest, conviction, or other civil requirements. Consequently, these images are typically high quality and uniform, resulting in consistency within and across agencies. Ten-print records are used in civil applications such as background checks and to determine immigration status; they are also used in the intelligence arena to identify known or suspected terrorists, and in the criminal justice system to identify arrestees and perpetrators of crime.

Figure 2. Example of a latent print.

Modern AFIS systems use automatic methods to encode a new ten-print image and search it against existing ten-print records, facilitating fast and accurate searches.

By contrast, a "latent print" (Figure 2) refers to a transferred impression of friction ridge detail, and includes those fingerprints recovered in whole or in part from a crime scene or other item of evidence. The highly variable characteristics of latent prints complicate the identification process. Latent prints are often incomplete or smudged, as they are usually created in an uncontrolled environment, thus reducing their quality and uniformity. Further chemical and physical processing and photographing or scanning of latent prints may be necessary to visualize the latent print before a search can be run using the image. When fingerprint examiners perform a latent print AFIS search, they mark up the features that can be identified on the image. This process is inherently subjective, and may be vulnerable to human error, especially in cases where the latent print is distorted or smudged or when the examiner is improperly trained or rushed.

Developers of AFIS software have independently created different methods for encoding the features of latent prints and algorithms for comparing latent prints to previously recorded ten-print records. These different approaches have made it difficult for the encoding features used in one vendor system to be used by another vendor system, resulting in insufficient interoperability among different AFIS systems to meet law enforcement agency needs.

**Identification of Missing Persons**

Fingerprints can be useful in identifying missing persons or victims of mass fatalities. Unidentified, burned, or fragmented fingerprints can sometimes be captured and used to make identifications. In some cases, medical examiners and coroners may be able to obtain ten-print records from bodies found at the scene of a crime or recovered from an accident. In other instances, latent searches can be performed on fragmented prints to assist with identification.

Modern AFIS systems create electronic data that, if standardized, can allow for interoperability. "Interoperability" is the ability of two or more networks, systems, devices, applications, or components to use standardized encoding to seamlessly and electronically share information on demand and as authorized, without special effort and without loss of accuracy. Interoperability is essential for law enforcement agencies to effectively investigate crimes and improve their ability to solve more crimes than stand-alone systems because it allows law enforcement agencies to search latent fingerprints against sets of ten-print records beyond those that are contained in their own databases within their own jurisdictions. While data sharing is a major component of interoperability, it is not the only criterion that is required for making interoperability possible. True interoperability requires technical compatibility, network connectivity, proper governance, and performance testing and training. It can be achieved only when an examiner can encode a latent fingerprint once and search the desired AFIS databases seamlessly, in accordance with relevant privacy laws, without having to re-encode the print or compromise search accuracy. Latent fingerprint examiners should be able to use the same machine for all of their searches and should not have to rely on manual re-encoding of images. This ability requires an integrated network of AFIS systems nationally that enables the exchange of fingerprint records to occur on demand as necessary.

Progress toward achieving interoperability has been made to date in each of these areas as follows:

- *Technical compatibility* has been mostly achieved for ten-print searches and is advancing with respect to latent fingerprint searches through the adoption of a common feature format, the Extended Feature Set (EFS), and the development of a transaction standard, the Latent Interoperability Transmission Specification (LITS).[2] The Federal Bureau of Investigation (FBI) introduced the Next Generation Identification (NGI) system (described at right), which has incorporated the EFS as the standard submission format.[3]
- *Network connectivity* and data sharing at present, are relatively limited. State agencies have direct access to NGI via the Criminal Justice Information Services (CJIS) Wide Area Network (WAN), a network designated specifically for transmission of extensive law enforcement information including AFIS searches of FBI records and some local agencies may search NGI without going through their states. Efforts are being explored to connect state agencies to one another and to connect local agencies to NGI through their State Identification Bureau (SIB) for latent searches.
- *Proper governance* has been facilitated through interagency coordination and guidelines on the process for developing agreements between agencies and recommended language for the agreements. A few localities have set up regional task groups to improveinteroperability within their jurisdictions, but many jurisdictions have not yet pursued such agreements.
- *Performance testing and training* has improved through the development of standardized guidelines for feature markup and the introduction of an online EFS training tool that allows examiners and AFIS vendors to explore the functionality of EFS.[4] NIST conducted standardized AFIS performance tests in 2011 and 2012 to assess the accuracy of systems using interoperable features, but tests have not been revisited since 2012.[5]

Despite these efforts, several core challenges to achieving true latent fingerprint interoperability remain. Few agencies have upgraded to systems that comply with the recently developed file format and transmission standards. In the interim, latent fingerprint examiners often forgo additional searches or must manually re-encode their prints on co-located workstations to perform searches on other jurisdictions' systems because they do not employ the same vendor to administer their AFIS systems. The re-encoding process takes time and requires additional personnel training. If not re-encoded properly, small differences can lead to a missed identification. Given the high volume of latent fingerprint examinations, the time burden resulting from re-encoding dissuades investigators from searching other databases in all but the most high-profile cases. Yet when additional searches are not performed, fewer matches are made, meaning that relevant suspects may go unidentified. The accuracy of image-only searches, in which the AFIS encodes the features, has increased greatly, but agencies' AFIS systems often do not provide a means for these transactions to be exchanged among agencies.

While re-encoding enables limited technical interoperability, seamless data sharing across networks or systems can occur only after adoption of information sharing policies between agencies and the integration of software to support the exchange of data. Absent these agreements or memoranda of understanding (MOUs) and technical infrastructure, remote searching cannot occur and even partial interoperability cannot exist.

**Federal Bureau of Investigation: Next Generation Identification**

The FBI's NGI allows access to the Repository of Individuals of Special Concern (RISC), a palm print repository, and access to other Federal biometric databases. NGI's fingerprint friction ridge feature encoders and matchers represent the latest generation of recognition technology, and the accuracy and speed of the system achieve optimal levels. Additionally, several of the factors that previously limited searches of the Integrated Automated Fingerprint Identification System (IAFIS) (*e.g.*, complex and proprietary encoding methods) have been eliminated or substantially mitigated.

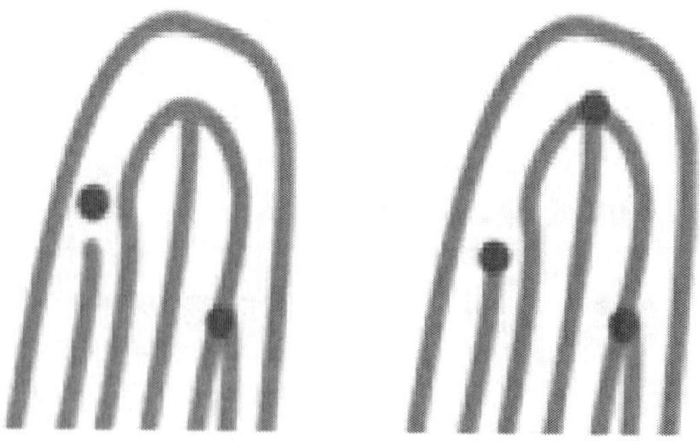

Figure 3. Encoding latent prints for two different AFIS systems by different vendors.

## Case for Latent Print Interoperability

In 2012, 53.2% of violent crimes in the United States went unsolved.[6] While it is not possible to predict what portion of these crimes might have been solved with a truly interoperable latent AFIS network, interoperability would certainly have aided investigations for some of them. Further, interoperability would help with investigations of unsolved non-violent crimes where latent prints have been recovered and where individuals have been criminally active in other jurisdictions. AFIS interoperability, if fully implemented for latent prints, would also improve the speed and efficiency of broader searches.

Many cross-jurisdictional searches have led to the identification of individuals that would have been missed if the law enforcement agency had searched only records contained in its own AFIS. Latent print examiners investigating a 2008 murder in Detroit, Michigan, searched both the Michigan State AFIS and the FBI IAFIS, but neither system generated a candidate suspect. In 2010, Michigan reached out to the Department of Homeland Security's United States Visitor and Immigrant Status Indicator Technology (US-VISIT) and sent a printed copy of the latent print. US-VISIT identified a potential match to the print that was not on file in either the Michigan State database or IAFIS. This ability to search multiple databases, which was facilitated by data sharing agreements, allowed the Detroit police to identify a suspect in the case. Remote searches have also led to the identification of suspects affiliated with Al-Qaeda, suggesting that interoperability would support both public safety and national security efforts.

The current lack of full interoperability has allowed criminals to remain free, and in some cases, commit additional crimes. An example that dates back to when the State connectivity to IAFIS was not complete highlights how these gaps challenge public safety efforts. In the 2002 "DC sniper" case, a series of shootings in the Washington, D.C. metropolitan area resulted in ten fatalities and three critical injuries. A month prior to the shootings, the scene of a homicide and robbery at a liquor store in Alabama produced fingerprints that were searched against the State AFIS with no match. At the time of the crime, the Alabama laboratory had not implemented a connection to IAFIS, nor had it forwarded the unknown latent prints from the case to the FBI. Nearly a month later, the FBI requested the prints following a call from a tip line indicating that the suspected sniper may have been connected to the Alabama crime. The latent prints from Alabama matched prints known to belong to an individual in the Immigration and Naturalization Service (INS) database. This INS search helped lead to the identification and subsequent apprehension of two suspects in the case, but the delay in identification, caused in part by the lack of system interoperability, contributed to the loss of life.

Since then, the FBI's NGI system has been improved and intentionally designed to provide state and local law enforcement agencies with the opportunity to conduct a latent fingerprint search by submitting additional information such as photographs, palm prints, and iris images.

## Historical Background of Fingerprinting and the Development of AFIS

Prior to the 1970s, latent fingerprint identification units within law enforcement agencies relied on a manual and laborious process to search latent fingerprints against ten-print records using the Henry Classification System.[7] This system of visually matching card-cataloged paper records to latent print features of ten-print records was time consuming and often did not result in an identification.

Individual state and local jurisdictions began purchasing AFIS systems in the 1970s and 1980s from various private vendors. Vendors capitalized on the fact that AFIS systems improved the likelihood of identification and dramatically reduced the amount of time necessary to identify or exclude a subject by searching against a database of electronic ten-print records of arrested and convicted offenders.[8] AFIS systems also allowed examiners to search both latent and ten-print prints against a greater quantity of ten-print records contained in an AFIS's computerized database.

Each vendor developed its own software and searching algorithms, which, over time, resulted in a stove-piped system of AFIS systems procured throughout the country. AFIS procurements were not coordinated among even geographically close law enforcement jurisdictions, and states did not require local agencies to purchase compatible systems. Independent purchasing decisions by agencies from the three major vendors and several smaller vendors created a national system where neighboring jurisdictions and even partners within the same state had vastly different systems. These stove-piped procurements resulted in an inability to automatically share electronic fingerprint data between state and local jurisdictions, even among agencies in geographic proximity and, more importantly, with a shared criminal pool. This lack of interoperability is in large part due to the lack of industry standards, which led to variability of vendor specifications and a lack of an integrated network supporting connectivity between two or more jurisdictions. These issues continue to impede crime-solving efforts, and while AFIS systems are interoperable for ten-print record searches, latent fingerprint interoperability has not been achieved because of the proprietary nature of the encoding schemes integrated into the different AFIS systems.

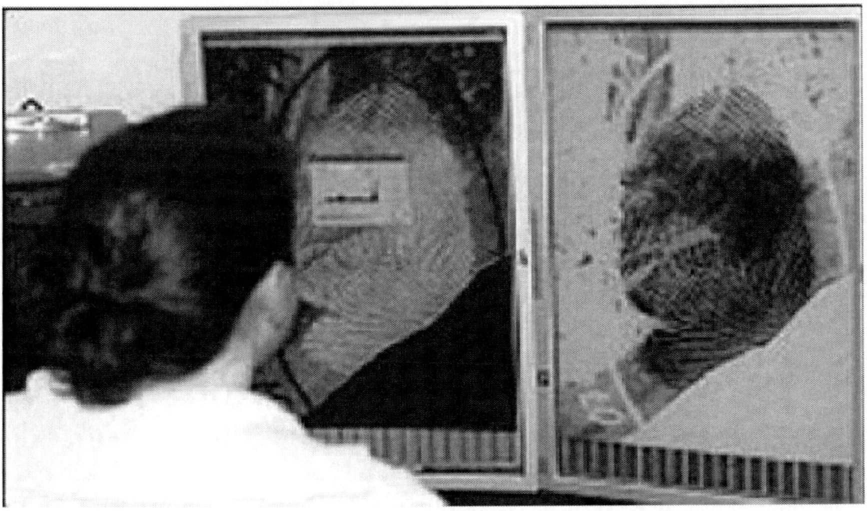

Figure 4. Latent print examiner. Source: FBI, "Houston Cold Case Solved," October 2011.

Efforts in the 1990s supported the development of standards that would enable state systems to communicate with the FBI's IAFIS, the system that preceded NGI. IAFIS became operational in 1999 and provided a national system that enabled ten-print records to be shared with the FBI. Through IAFIS, all State-level law enforcement agencies have the capability to submit civil, criminal, and latent prints to the FBI via the CJIS WAN, a collection of Virtual Private Network (VPN) links and near point-to-point T-1 and higher class data lines connecting the FBI CJIS Data Center in West Virginia to selected points throughout the United States and Canada."[9]

In creating IAFIS, the FBI had to accommodate technical limitations at the time of development. While prints could be submitted directly for searching against the system, as is currently done with ten-print searching, automated feature extraction algorithms were not advanced enough to achieve sufficient accuracy. As a result, latent examiners had to manually re-encode shared prints. This re-encoding process had the potential to introduce systematic human errors when the FBI's feature definitions differed from those required by the state or local AFIS. It also decreased the likelihood an FBI search would be made because to perform these extra searches, examiners had to re-encode images to comply with the IAFIS submission requirements.[10]

To standardize submissions to IAFIS (and now to NGI), the FBI required compliance to EBTS.[11] EBTS is based on a standard developed by the American National Standards Institute (ANSI) and the Information Technology Laboratory of NIST (ANSI/NIST-ITL 1-2000). These standards included specifications on image resolution, common field names, and how to include personal information and details on why the fingerprint record was created. EFS was incorporated into ANSI/NIST-ITL in 2011 and into EBTS in 2012. More recently, NIST has developed the LITS specification to specify how EFS features are to be organized and marked up for interoperable data exchange between latent AFIS systems. LITS is fully compatible with the FBI's EBTS.[12]

As more biometric matching capabilities became available, the FBI developed the NGI system, which expanded search capabilities and improved speed and accuracy. A recent major development in NGI was the implementation of a palm-matching capability which significantly extends the ability of the examiner to identify latent prints. NGI's adoption in 2013 has provided the latent fingerprinting community with an additional incentive to improve interoperability in order to benefit from the FBI's system upgrades.

## EXISTING MECHANISMS FOR FINGERPRINT SEARCH COORDINATION AND SHARING IN A TIERED AND FRAGMENTED SYSTEM

The current limited landscape of AFIS interoperability highlights and reinforces the need for interoperability. Multiple tiers of law enforcement agencies perform uncoordinated searches, inhibiting criminal investigations in contrast to an interoperable environment (Figures 5 and 6).

Although built primarily by three major vendors, the majority of the estimated 600 disparate AFIS systems at the Federal, state, and local levels are not interoperable from a technical or governance standpoint. As a result, data sharing is infrequent and case-specific, rather than routine. The sharing that does occur happens primarily through siloed channels in an established hierarchy from local to state and state to the FBI's NGI.

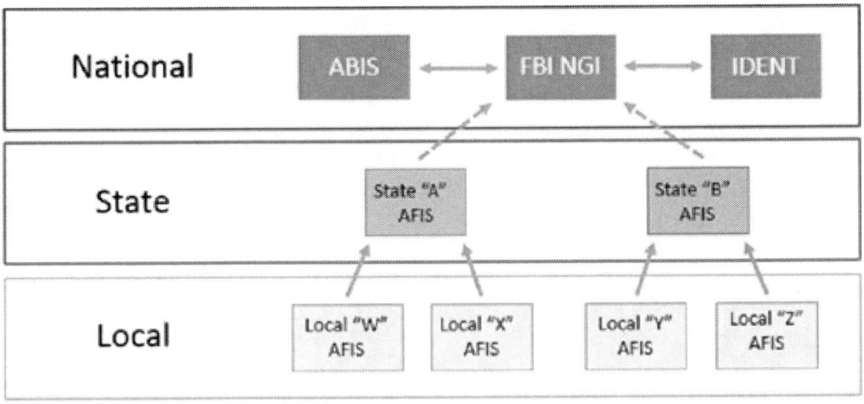


Figure 5. Current AFIS sharing environment.

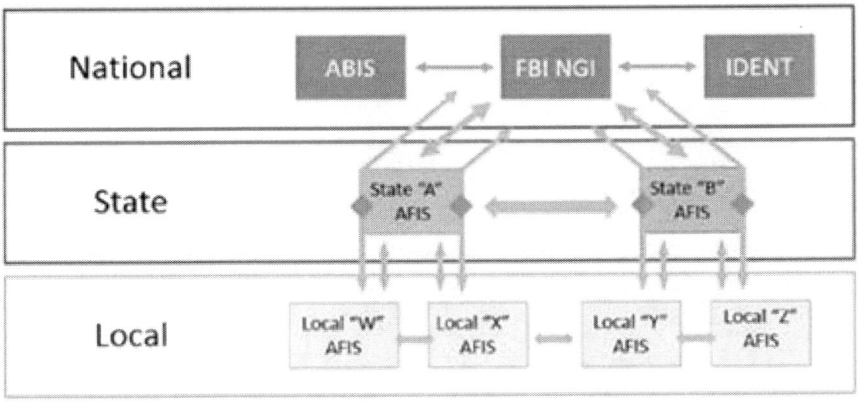


Figure 6. AFIS interoperable environment.

Local law enforcement agencies often search only their own AFIS, sometimes search their state's AFIS, and rarely share data with neighboring jurisdictions. At the next tier of the hierarchy, state law enforcement agencies rarely submit latent print searches to neighboring states (with the exception of some unique regional agreements). States report low levels of searching against the NGI or neighboring jurisdictions' AFIS databases, even for unsolved cases. While most crimes are solved at the local level and would not require broader sharing, there is some hesitancy to share when appropriate, in part because of limited adoption standards for compliant systems and outmoded examiner practices.

Table 1 highlights agencies' perceptions of their current data sharing levels, and their predicted level of sharing if all of the technical barriers were removed. Significant variability exists among state and local data retention laws and policies.[13] Therefore, innovative governance structures would also need to be introduced to harmonize data retention laws and facilitate data sharing while respecting civil liberties. Without these changes, agencies would have great difficulty conducting an independent search on another local or state AFIS.

**Table 1. Percentage of Latent Interoperable Searches by Select Jurisdiction**

| Geographic Area | % Searches Sent to FBI | | % Searches Sent to State | | % Searches Sent to Neighboring Jurisdictions | |
|---|---|---|---|---|---|---|
| | *Current* | *If Seamless* | *Current* | *If Seamless* | *Current* | *If Seamless* |
| *LOCAL AGENCIES* | | | | | | |
| Northern Virginia | 1% | 40–50% | 1% | 40–50% | 80–90% | Already seamless |
| *STATE AGENCIES* | | | | | | |
| Baltimore, Maryland* | 0% | 0% | — | — | 0% | 0% |
| Kansas City, Missouri | 10% | 20–30% | — | — | 0% | 20–30% |
| Michigan | 3–5% | 100% | — | — | Rarely | 100% |
| New York | 80% | 80% | — | — | 10% | 80% (All non-identificationson State AFIS) |
| *REGIONAL AGENCIES* | | | | | | |
| Portland, Oregon(Western Identification Network) | 15% | 20–25% | 100% | 100% | All non-identifica-tionson State AFIS | All non-identifications on State AFIS |
| Las Vegas, NV (local AFIS & WIN) | Very rare | All non-identifications on local AFIS | All non-identifica-tions on local AFIS | All non-identifica-tions on local AFIS | All non-identifica-tions on local AFIS | All non-identifications on local AFIS |

Source: Noblis, Inc. Latent Print Interoperability: State and Local Perspectives, April 2, 2012.

* Representatives from Baltimore indicated that, given the current latent print examiner staffing level, they could not use interoperability even if it were available.

## Sharing at the Local Level

Many cities, counties, and metropolitan areas have independent AFIS systems that are not interoperable with their own State systems. Most of the local agencies that had their own AFIS databases and responded to a 2013–2014 National Institute of Justice (NIJ) survey on latent fingerprint interoperability reported that they frequently searched their state agencies' data.[14] However, according to state agencies, local agencies in only about half the states forward their unsolved latent prints to state agencies.[15] The survey also found that only 15 states reported that local law enforcement agencies forwarded all unsolved latent prints to the state agency, which is likely due to policy limitations.[16] While some local AFIS systems are directly connected to the state AFIS, a substantial proportion are not, especially those developed by different vendors.[17] This lack of connectivity weakens the investigative power of the local law enforcement agency, because most state AFIS systems do not contain the sum of all local AFIS data within the state. As a result, suspects who operate in multiple counties may avoid identification.

If local agencies routinely shared fingerprint data among themselves, some of the concerns about the comprehensiveness of state AFIS records would be assuaged. But local-to-local sharing is infrequent and especially difficult when neighboring jurisdictions cross state lines. Searching neighboring AFIS systems is conducted ad hoc and depends on personal relationships among colleagues. Only 19% of responding local law enforcement agencies reported frequent searching of their neighboring jurisdictions.[18] When cross-jurisdictional searches occur, many local agencies prioritize searches by severity of the offense, and 85% of agencies in the NIJ survey who reported prioritizing latent searches based on "criteria associated with the criminal offense under investigation," also reported that "major crimes are prioritized regardless of the jurisdiction" of origin.[19] Law enforcement agencies face similar challenges partnering with other local jurisdictions governing the same geographic region. Some larger municipalities with their own independent AFIS do not have the same AFIS vendor as the county agency and must re-encode latent prints on a separate workstation to search the county AFIS.

## State and Local Access to Next Generation Identification

Every State in the Nation has direct access to IAFIS through the CJIS WAN, a dedicated network connecting state and local agencies to the FBI.

Further, about 50 local law enforcement agencies are able to directly submit searches through Direct Local Connectivity (DLC). Despite this connectivity, when asked if they ever searched latent prints on IAFIS or NGI, five state agencies reported that they had not, though 31 states responded that they searched IAFIS routinely.[20]

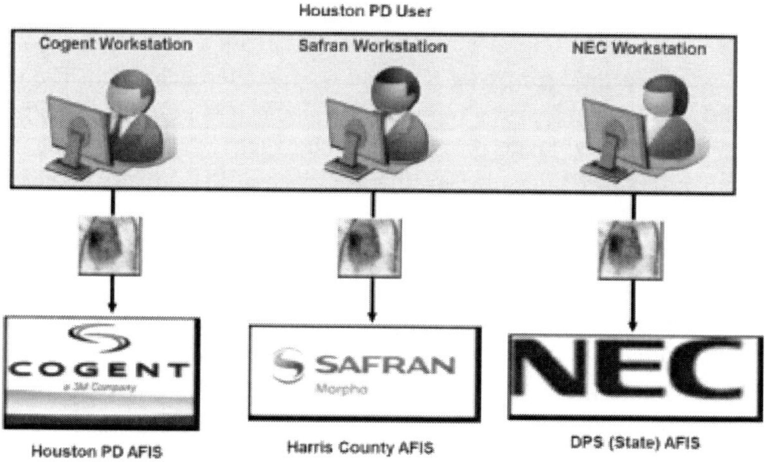


**Barriers to Sharing and Interoperability in Houston, Texas**

The city of Houston is the largest city in Texas and the fourth-largest in the country. Houston is within Harris County, which is the most populous county in the State and the third-most populous county in the country.

Until recently, Houston and Harris County maintained separate AFIS databases administered by different vendors that were incompatible. The State of Texas also maintained its own AFIS database administered by a third (different) proprietary vendor. For a latent print examiner within the Houston Police Department to access the Harris County AFIS database and the Texas State AFIS database, the examiner had to physically access different workstations, each requiring a different method to re-encode the latent print before conducting the searches. Now latent fingerprint examiners can use the Universal Latent Workstations or export a compliant transaction from their proprietary AFIS that can be submitted to the FBI, through the Texas Department of Public Safety, without re-encoding.

This access is primarily through a state law enforcement agency, using separate Universal Latent Workstations (ULWs), which use free software provided by the FBI that prepares fingerprint data to be sent to the FBI's secure network CJIS-WAN. Among local agencies that reported submitting searches to IAFIS, about half reported submitting them directly and the other half reported submitting them through their state law enforcement agencies.[21] When IAFIS was first implemented the state-level daily query limit for latent searches exceeded states' current utilization of the system. In fiscal year 2014, 13,000 remote searches were performed on average per month. The submission rate was influenced by workflow issues that arise from examiners needing to re-encode minutiae because of lack of AFIS interoperability. The need to re-encode makes searching NGI time-consuming and requires additional training. State examiners may also be less likely to pass along local searches to the NGI unless they are relevant to state law enforcement agency investigations.

## Regional Interoperability Agreements

According to the NIJ survey, the majority of states (77%) receive latent requests from law enforcement agencies with an AFIS from outside of their state, but just over half of all states described these requests as being routine.[22] Some regions have formalized these sharing relationships to facilitate routine cross-border sharing. Other states have developed bilateral sharing agreements, though these agreements fall short of specifying interoperability (see box on the Western Identification Network below). Such agreements often do not support reciprocal sharing and can be invalidated if one agency switches to another AFIS vendor or upgrades to a newer version of its existing vendor's software.

**Western Identification Network**

The Western Identification Network (WIN) is the most comprehensive regional sharing agreement. Eight western states share a common Automated Biometric Identification System (ABIS) Service Bureau and maintain reciprocal search agreements with the California Department of Justice, Las Vegas in Nevada, and Riverside-San Bernardino counties in California.

State WIN agencies that are part of the central service bureau and participating interface agencies have access to more than 30 million fingerprint records and search WIN's AFIS without re-encoding searches. Local agencies can access WIN via a WIN workstation or through the standards-based Universal Latent Workstation (ULW) and ten-print submissions from any authorized workstation. In 2012, there were 7,000 daily ten-print submissions and 500 latent print submissions to WIN.

For agencies that have switched or plan to switch AFIS vendors, reciprocal search agreements will remain in place and allow agencies to either maintain a WIN workstation or use the recently adopted EFS specification accessed via the ULW. Regional connectivity continues to be supported by maintaining a dedicated WIN workstation or implementing ULW software or rapid-standardsbased ten-print searches. The WIN workstation supports full system capability (e.g., registration, archive access, latent case management, etc.), while the ULW supports only searches of WIN's database.

Regional systems such as WIN and the Northern Virginia Regional Identification System (NOVARIS) have successfully implemented data sharing agreements. These arrangements reveal the benefits and difficulties of full-scale interoperability. Largely, the interactions among the members in these systems have developed in areas where law enforcement agencies shared common vendor or co-located workstations. Unique contractual agreements—usually an MOU that has been developed with the help of the incumbent vendor—have formalized data sharing. If an agency within the network were to change vendors, the data sharing mechanism would likely be broken, because most of these agencies have not yet implemented the data sharing standards necessary for interoperability. Smaller regional interoperability agreements have also been implemented across the United States.[23] As with larger sharing networks, these can be jeopardized if an agency changes its vendor.

## Fingerprint Sharing among Federal Systems

At the Federal level, three agencies maintain AFIS systems that are semi-interoperable. The FBI's Next Generation Identification (NGI) system operates in parallel with national fingerprinting databases maintained by the

Department of Homeland Security (DHS) and the Department of Defense (DOD).

DHS maintains the Automated Biometric Identification System (ABIS) called IDENT, which is a program of the Office of Biometric Management (Figure 7).[24] IDENT is the central DHS-wide system for storage and processing of biometric and associated biographic information for national security, law enforcement, immigration and border management, intelligence, and other DHS mission-related functions. IDENT was originally developed in 1994 for the Immigration and Naturalization Service (INS). With the incorporation of the INS into DHS in 2002, the use of IDENT was expanded to be DHS's primary biometric identification system. Today, IDENT helps DHS meet many of its statutory requirements under the "Uniting and Strengthening America by Providing Appropriate Tools Required to Intercept and Obstruct Terrorism (USA PATRIOT) Act of 2001,[25] the Homeland Security Act of 2002,[26] the Enhanced Border Security and Visa Reform Act of 2002,[27] and the Intelligence Reform and Terrorism Prevention Act of 2004.[28] IDENT holds over 170 million individuals' fingerprints provided by DHS agencies such as Immigrations and Customs Enforcement, Customs and Border Protection, United States Citizenship and Immigration Services, Transportation Security Administration, and the United States Coast Guard, as well as other Federal agencies. In addition, IDENT contains over 220,000 latent prints.

Latent interoperability between NGI and IDENT is limited because DHS has not adopted EFS and therefore must re-encode latent prints received from NGI for search in IDENT to ensure the highest level of search accuracy. These re-encoding burdens restrict the number of searches between the two systems based on available manpower and time for manual throughput. In contrast to most cross-jurisdictional submissions where latent re-encoding is required, DHS will re-encode prints upon receipt, as opposed to requiring the submitting agency to re-encode the print to ensure the highest level of search accuracy within IDENT. The need to re-encode could be eliminated if DHS were to adopt the EFS specification.

The Biometrics Identity Management Activity (BIMA), which falls under the Defense Forensics and Biometrics Agency in DOD, operates the authoritative multimodal biometric database known as DOD ABIS. What began in 2004 as a suite of technologies for securing access to military installations is now recognized by warfighters as an innovative tool for identifying and tracking known or suspected terrorists. For example, in Iraq and Afghanistan, the data contained in DOD ABIS helped counter the threat of improvised explosive devices by tracing latent fingerprints back to those who made or detonated these weapons. In 2009, DOD extended the capabilities of DOD ABIS to include face and iris matching, which, when combined with fingerprint data, reduce response time and human intervention and increase identification accuracy. The DOD ABIS database currently contains more than 12.3 million biometric files and over 300,000 latent prints collected from military theaters of operation.

In support of homeland security, BIMA shares ABIS data with NGI and IDENT. As of December 2013, DOD ABIS achieved compliance with the FBI's standard for ten-print submissions, but has not yet done so for latent prints. In practice, DOD latent print submissions are encoded utilizing the FBI's ULW to ensure interoperability. NGI latent print submissions to DOD ABIS are performed by an internal translation system that converts NGI encoded prints into DOD-compliant submissions. DOD's latent prints are pushed to IDENT via the FBI where they must be manually re-encoded prior to searching. DHS notifies the DOD via email of any identifications made in IDENT to DOD latent prints.

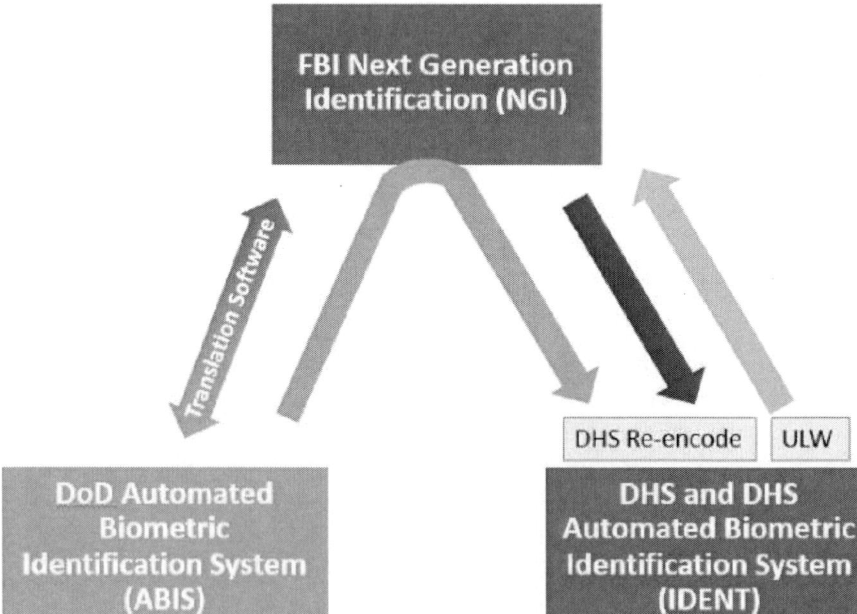


Figure 7. Interoperability among Federal agencies.

The connections between states and these Federal agencies primarily occur through NGI. While most states routinely search NGI, only six states have ever directly requested a latent print search from another Federal agency database, only one state is directly connected to other Federal systems, and three others connect through NGI.[29]

## Interoperability Pilot Programs

Several pilot programs seek to provide local law enforcement agencies with direct access to NGI and other Federal AFIS systems. By providing an interoperability solution to localities, the capabilities offered by Federal databases can be used more effectively for specific investigatory purposes. These pilots also serve to demonstrate the value of interoperability and mitigate challenges by identifying potential technical and operational solutions.

## Latent Interoperability Pilot Program

In an effort to expand the availability of latent fingerprint services of DHS's IDENT, DHS and the Texas Department of Public Safety (TXDPS) are developing a Letter of Intent that would permit TXDPS to electronically submit searches to IDENT. Once the agreement is finalized, TXDPS will be able to use existing connectivity provided by the CJIS-WAN to electronically search DHS's IDENT, as TXDPS now does with the ABIS and NGI. Current searches submitted to DHS by TXDPS are done so by non-electronic means. The pilot will help evaluate IDENT's expanded latent print services and its value to the greater law enforcement community at the local level, especially in jurisdictions along the U.S. border.

## Other National Pilots for Latent Interoperability

Recent initiatives have focused on introducing latent search capabilities to new users that might benefit from fingerprint identification. These initiatives have allowed the following users to have access to AFIS systems:

- Medical examiners and coroners (ME/C),
- Customs and border protection agents, and
- Patrol officers attempting to identify subjects with fingerprints of reduced quality.

**AFIS Searches at the Medical Examiner's Office**

A Harris County, Texas, morgue employee used a mobile scanner (and a traditional table-top model) in an attempt to identify and confirm the identification of decedents. The use of the mobile device greatly facilitated the fingerprint capture process, allowing the non-expert to scan all fingers.

After receiving the scanned prints, the Harris County Sheriff's Department AFIS database searched for a match and, if available, returned mug shots for identity confirmation. Several identifications against the unsolved latent fingerprint database maintained by Harris County were also obtained as a result of this project.

Mobile handheld fingerprint scanning devices with access to an AFIS database have made it possible for these users to incorporate fingerprint identification into their investigations and duties. Most of the data captured by these mobile devices is searched using an automated ten-print or two-print identification process. But some subjects, such as decedents and individuals whose fingerprints are collected in less than ideal conditions, can result in poor quality fingerprint images and require the ability to search using latent fingerprint encoding methods to enhance accuracy. Once these are properly encoded, multiple databases may need to be searched to identify a match.

In most cases, ME/C offices do not have direct authority to access a local jurisdiction's AFIS and must go through a law enforcement agency to perform a search, which impedes their ability to identify decedents. Several pilot programs have placed AFIS workstations and the proper governance in place in ME/C offices to allow them to search the ten-print records maintained at the local or state level.

To achieve interoperability and to take advantage of the information contained in different AFIS databases, changes in laws and policies governing access will need to be considered, and special provisions may need to be included in MOUs established between jurisdictions to allow for access by ME/C and other authorized persons outside law enforcement agencies.

## Recent Progress toward Interoperability

### *Technical Compatibility through the Establishment of Standards*

Many interoperability problems arose out of the variable and proprietary nature of individual agencies' AFIS system encoding specifications for latent prints. In particular, minutiae used for identification in one vendor's AFIS might be unreadable or irrelevant in another AFIS. NIST and the Committee to Define an Extended Fingerprint Feature Set worked extensively to establish a standard set of friction ridge features and minutiae that must be included to search other systems, as an interoperable AFIS exchange format. The EFS specification provides comprehensive and consistent definitions of minutiae for use in fingerprinting. It also specifies methods for encoding features found on fingerprints and palm prints as well as how to annotate the quality of the features.

In addition, NIST and ANSI specified which EFS features need to be incorporated to submit or receive a remote search regardless of which vendor made the systems. This vendor-neutral transaction standard is known as the Latent Interoperability Transmission Standard (LITS). Since incorporation of LITS into the fingerprint and biometric ANSI/NIST-ITL 1-2011 standard in 2011, there has been mixed success in adoption of the feature set and transmission standard by vendors because localities generally are not making it a requirement of the contractual agreements with their AFIS vendors.[30] Long-term procurement cycles also make it difficult to update all of the AFIS. In a 2013–2014 survey, the average state and the average local agency reported that their AFIS was upgraded or became operational in 2008.[31] Over 60% of states have reported that they plan to update their system or release a procurement announcement for a completely new AFIS in the next 3 years, which presents a unique opportunity to influence the procurement process.[32]

**EFS and LITS**

The EFS specification provides comprehensive and consistent definitions of fingerprint and palm print features for use in fingerprinting. This standard defines the features to be used in both ten-print and latent print searches. It refers to the fingerprint and palm print features incorporated into the ANSI/NIST-ITL standards.

The LITS specification describes what information is required for a latent print search transaction to occur across jurisdictions, regardless of originating and destination AFIS vendor. LITS is compatible with the FBI's EBTS.

### *Transition to Next Generation Identification*

The FBI's transition to the NGI incorporates functionality that will benefit the latent print examiner community. By implementing the recently adopted standards, NGI will create a more formal channel for local agencies to submit searches to the system and will add palm prints and other biometric identifiers such as facial and iris recognition into records, which will likely increase search and identification accuracy. In early 2013, as part of the third increment of NGI implementation, latent print functionality was migrated from IAFIS to NGI. The new system allows for faster searches, a simplified and standardized encoding process, and the ability to prioritize the most pressing searches.

NGI permits latent print examiners to search against the entire criminal database or a particular subset of records such as the Repository of Individuals of Special Concern (RISC).[33] For example, law enforcement officials can submit a search to the RISC from remote locations, such as during a traffic stop, with a small handheld device to quickly identify high-interest individuals. Officers are notified of a match probability for each potential candidate in the RISC, using a stoplight color-coded system, (red, yellow, and green). Searches submitted to RISC are also cascaded (automatically searched) against the Unsolved Latent File (ULF), a group of unsolved cases,[34] which may help identify suspects in older, unsolved cases. Unlike ten-prints in the RISC system, the ULF search does not immediately return the personal information related to a match. Instead, RISC notifies the examiner and forwards potential matches for verification. RISC contains only a subset of NGI's records.

Beyond some of the technical limitations that still remain with NGI as law enforcement agencies transition to standards compliant systems, the overall challenge with the transition to NGI is the limited capacity of latent examiners to run an additional search in the NGI system or compare the resulting candidate lists. A number of states do not routinely submit searches to NGI due to time and resource constraints.

### *Improving Latent Print Algorithms*

While vendor competition has hindered collaboration and cross-jurisdictional compatibility, it has encouraged vendors to strive for more accurate search algorithms. Advances in latent print identification algorithms have allowed latent print examiners to identify more suspects, especially with poor quality prints due to increased accuracy of image processing and matching algorithms. Vendors are also starting to use more robust ensemble methods to combine multiple search algorithms to generate more precise candidate lists. Ongoing research is investigating methods for more effective candidate list management, including fusion algorithms to limit the number of candidates reviewed and merge candidate lists from multiple searches.

These recent technical advances have allowed the law enforcement community to work towards building an interoperable system capable of quickly and efficiently solving more crimes. Much more needs to be done to bring the system from a fragmented system of ad hoc sharing to one that is fully interoperable. The next section provides a path forward from the current landscape to an interoperable latent print system. Image processing algorithms that encode latent prints are approaching the accuracy of human examiners and soon may eliminate the need for human encoding in most cases, which would likely speed up the latent print search process. Improved encoding, while speeding up the submission process, does not eliminate the need for the examiner to compare candidate lists returned as a result of a particular search.

## PATH FORWARD: ESSENTIAL ELEMENTS FOR INTEROPERABILITY

An alternative to the current multiple AFIS databases would be a system whereby all law enforcement agencies in the United States submit every latent print to a central repository of biometric information, within the bounds of standardized privacy laws. Ideally, this system would be a single interface that accesses all law enforcement systems, allowing users to select to run a search against a local, state, or national database, when appropriate. Such a system could provide seamless technical compatibility, network connectivity, a comprehensive governance framework that ensures secure transfers and respects privacy laws, and rigorous quality assurance mechanisms for both technology and staff that use and manage the system.

Maintaining a centralized system would likely be infeasible given the cost of procurement and retraining and the logistical concerns associated with connecting every law enforcement agency. The elements necessary to maintain an effective single system are the same as those in an environment of distinct interoperable systems.[35] Additionally, education and outreach are necessary to promote advances in these other areas. The Federal Government and state and local agencies have made varying progress on implementing these elements. The following sections define the requisites for each component, enumerate progress that has been made toward incorporating each element, identify outstanding challenges, and provide recommendations to address these challenges.

## Technical Compatibility

Interoperability depends on the technical compatibility of electronic data submitted and received between two or more agencies' software systems. In the context of latent print searches, compatibility requires consistent fingerprint feature definitions and designation of a minimum set of features that must be encoded on the digital image to conduct a latent print search. These "feature sets" allow for latent prints to be compared to both ten-print records and other latent print records within an AFIS.

The AFIS market has always been competitive, allowing multiple vendors to successfully market their products to different law enforcement agencies. Engineers from different vendors built systems that had varying algorithms to define requirements for identifying and weighting the features of a latent fingerprint image. In response to the requirements specified in law enforcement agencies' Requests for Proposals (RFPs), engineers prioritized accuracy and ease of use within a single AFIS, instead of compatibility with other AFIS. Vendors, furthermore, often did not include legacy compatibility, thus making newer AFIS systems incompatible with older versions from the same vendor.

## Recent Progress on Compatibility and Remaining Challenges

Recognizing the need for compatibility, the Committee to Define an Extended Fingerprint Feature Set,[36] composed of a broad spectrum of stakeholders, including law enforcement officials, latent fingerprint examiners, academics, and engineers from major AFIS vendors, set out to establish a features encoding standard to be incorporated in the ANSI/NIST-ITL 1-2011 standard, which was updated in 2013. The result was the EFS, which created standardized minutiae and friction ridge feature definitions and standardized data field names and indexing. This updated standard defines the features to be used in both ten-print and latent print searches and is consistent with the FBI's broader biometric standard, EBTS. EBTS has been incorporated in major systems including NGI and DOD's ABIS.[37] NIST testing of the EFS has demonstrated that it provides the basis for a common set of features that all major vendors can use.[38]

Building on this standard, the NIST Law Enforcement Standards Office, now the Office of Special Programs Forensic Science Program, developed the LITS document that describes what information is required for a search across jurisdictions.[39] LITS requires that latent print submissions include transactional meta-data, the latent print image, and minutiae data. By incorporating LITS directly into the local and state AFIS, law enforcement agencies will reduce the need to encode latent prints on multiple workstations. In doing so, it will remove many of the technical and workflow barriers that limit fingerprint data sharing and interoperability while ensuring compatibility with EBTS and EFS standards previously established.

The LITS designates two different search profiles that are required across all interoperable systems: the image-only search and the quick minutiae search.[40] These profiles are vendor-neutral and can be used across systems that have adopted the LITS or are compatible with EBTS (NGI). Additional information can be encoded for use by specific vendors or future systems by submitting additional markup details through optional profiles (such as the detailed profile markup shown in Table 2). The additional profiles may allow examiners to increase their accuracy by encoding additional features, including vendor-specific features. Examiners have to weigh the tradeoff between time spent encoding additional features and time to perform comparisons or search additional databases.

Incorporation of the LITS is ongoing and vendors are beginning to build systems that natively use the EFS features and profiles defined by LITS. To support the adoption of LITS-compliant systems, NIST has developed writing guidelines for agencies to use in preparing their RFPs.[41] As of September 2014, few law enforcement agencies had purchased LITS-compliant systems because upgrades are both costly and require substantial training of print examiner staff. Of those law enforcement agencies who responded to the recent NIJ survey and reported having upgraded their systems, the last upgrade occurred on average 6 years ago for State agencies and 5 years ago for local agencies, prior to the adoption of the LITS specification.[42] While over two-thirds of responding state agencies reported that they expected a major upgrade of their systems within the next 3 years,[43] only 35% of responding local agencies expected a major upgrade within that timeframe.

In the interim, a subset of law enforcement agencies are using the ULW to support their interoperability needs. The ULW software is freeware provided by the FBI that can run on most computers and that allows the examiner to encode using interoperable EFS features. Most states currently receive all search results from the FBI through the ULW.[44]

**Table 2. EFS profiles required by LITS and an example of an additional EFS profile supported by LITS**

| LITS Required | | LITS Optional |
|---|---|---|
| Latent Image-Only Search | Quick Minutiae Search | Detailed Markup Profile |
| 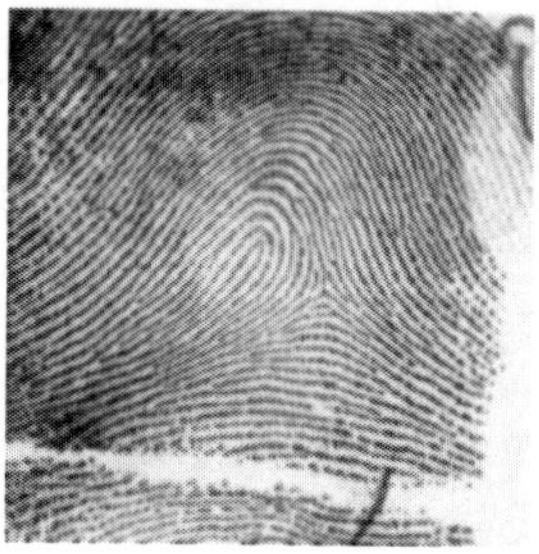 | 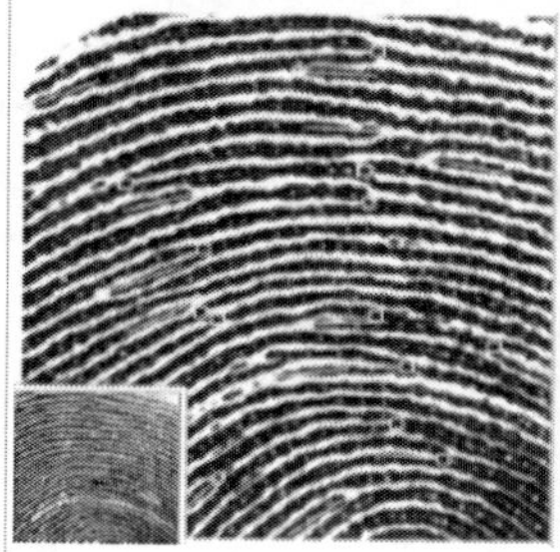 | 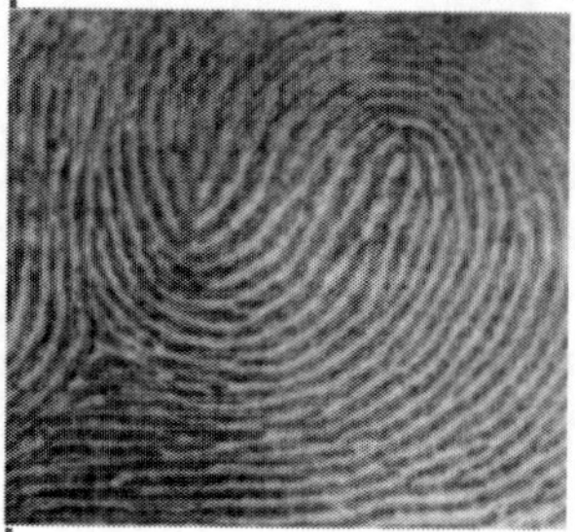 |
| The image is properly cropped and submitted without markup or annotations. | The latent print examiner must mark region of interest, minutiae, cores, deltas, pattern, and orientation. | Ridge Flow Map is one of the features included as part of the detailed profile. |

Source: Austin Hicklin, Standardizing a More Complete Set of Fingerprint Features (Noblis, Inc., 2007), prepared for the Committee to Define an Extended Fingerprint Feature Set. Screen shot from EFS Training Tool, www.nist.gov/forensics/EFSTrainingTool/FundamentalAFISSearching/MarkingMinutiae.html.

Many vendors have not incorporated export or translation functionality to the ULW. Without this functionality, latent examiners must manually re-encode prints to submit them to NGI. Half of states currently re-encode prints prior to submitting a search to the FBI.[45]

The existence of biometric standards alone is not enough to demonstrate that an AFIS system meets the technical requirements specified in the standards. Conformance testing is also necessary to ensure LITS compliance. Information on an AFIS system's conformance to a particular standard can provide an efficient method of conveying information on the product's suitability. Conformance testing, which captures the technical description of a specification and measures whether the specification has been faithfully implemented, has yet to be developed for LITS and EFS. Future tests should evaluate the data structure (syntactical conformance), the data content and relationships between fields (morphological conformance), and whether the data represent the parent biometric data (semantic conformance).

## Path Forward for Compatibility and Compliance

The Federal Government could take several steps to help speed the adoption of LITS-compliant AFIS systems and ensure compliance to newly implemented standards. First, the Federal Government could set a clear example by ensuring that all Federal AFIS systems are made fully LITS compliant without the use of an additional workstation within 3 years. Next, the Federal Government could provide support to state and local agencies through existing grant programs for achieving LITS compliance.

Federal funds could be made available to state and local agencies for AFIS procurement or upgrades, provided the systems are LITS-compliant. Finally, NIST should ensure that the recently adopted standards are appropriate once implemented and are reviewed in 3 years. NIST could also ensure that updates to the ANSI/NIST ITL 1-2011 standard are backward-compatible and provide guidance on how to improve interoperability with legacy systems, while agencies upgrade their systems in the interim.

**Recommended Criteria for Conformance Testing Programs**

- A common application programming interface with detailed definitions and requirements to evaluate each component of the AFIS technology
- A neutral third-party organization capable of administering the test, maintaining the secrecy and integrity of the test data and AFIS technology under test, and analyzing and reporting the results
- The implementation of an open-set (one-to-many) identification testing protocol
- A sufficiently large set of sequestered test data—latent prints and exemplars—that are operationally representative of data the system would be expected to process with respect to quality, distribution of impression types, and other factors
- Feature markup in accordance with the LITS and EFS profiles and guidelines produced both with and without the benefit of the associated exemplars as reference sample
- Single- and multi-exemplar sets per enrolled subject

To validate LITS-compliance, NIST should develop conformance testing standards to assess the incorporation of the ANSI/NIST-ILT, EFS, EBTS, and LITS standards and specifications into new AFIS systems. NIST could either develop a conformance testing program or fund independent programs that assess an AFIS's compliance to these standards. (See text box on Recommended Criteria for Conformance Testing Programs.)

## Network Connectivity

The success of law enforcement agencies' coordinated efforts to apprehend suspects of crime across jurisdictional lines depends on the connectivity of different AFIS systems through established networks. Increased connectivity would help overcome the limitations to electronic data sharing that currently exist among local-to-local communications and state-to-state communications. Current methods of sharing fingerprint image data span from transmission over electronic secure networks to delivering CDs through the United States Postal Service.

These gaps arose for many of the same reasons that compatibility issues arose. Individual jurisdictions procured AFIS systems from different vendors without consideration of connectivity with other systems within close geographic proximity. Varying privacy laws also prevented states from creating an interconnected network. Where connectivity does exist, it tends to occur in situations where agencies possess AFIS systems from the same vendor. Vendors, in fact, would have a disincentive for local connectivity because it could result in economies of scale that could have a negative financial impact on the industry.

NGI access aside, fewer than half of states reported having a direct connection to another law enforcement agency's AFIS system, even though over half of state and local law enforcement agencies reported being part of an AFIS network that contained more than one agency, This suggests that many networks do not permit interoperable data sharing.[46]

## Recent Progress on Network Connectivity and Remaining Challenges

Three main networks facilitate the majority of latent print communications among law enforcement agencies: the CJIS-WAN, the International Justice and Public Safety Network (Nlets), and the Law Enforcement Enterprise Portal (LEEP), formerly the Law Enforcement Online network.

Connectivity between the state AFIS systems and NGI is made possible through the CJIS-WAN. Since its adoption the CJIS-WAN has allowed the submission of both ten-prints and latent prints from states to NGI. Most local agencies do not have direct access to NGI, and therefore must contact their state agencies for access. Local agencies that have the ability to directly submit searches against NGI must submit them through LEEP.

Nlets, a not-for-profit organization managed by state law enforcement agencies, runs a network used to share a wide array of criminal justice and law enforcement information, domestically and internationally. The WIN network and New Hampshire, Maine, and Vermont share interstate latent print information through Nlets.

The FBI's LEEP is an online network used by law enforcement agencies to support investigative operations, to send notifications and alerts, and to provide an avenue to remotely access other law enforcement and intelligence systems and resources. It is not intended to support latent AFIS searches, but developers have created a software wrapper to send emails with attached latent images via LEEP.

None of these systems has the capacity to fully support interoperability among agencies performing latent print searches. The current connectivity system landscape creates major hurdles for connecting agencies both horizontally (locality-to-locality or state-to-state) and vertically (locality-to-state). The CJIS-WAN would require major redesign and redeployment to support extensive local-to-local links, but it may be an alternative for linking states to other states. State agencies are already connected to the FBI through CJIS-WAN and the CJIS Advisory Policy Board has already approved the use of CJIS-WAN to route messages between states. Local agencies may only connect to Nlets through their states' connections, limiting local agencies' ability to make use of the system, and the current system does not support interstate local sharing.

Beginning in June 2017, CJIS will no longer allow local law enforcement agencies to directly access NGI. Direct Latent Connectivity (DLC) services had been extended to select local law enforcement agencies as an interim solution while states developed the programming and telecommunications infrastructure to support local connectivity for latent services. Advances in technical compatibility are obviating the need for the DLC services. To ensure the continuity of access for local agencies, connected local agencies were required to work with states to develop and submit a formal Transition Plan to CJIS by December 2014. Once DLC is no longer supported, SIBs will steward additional local connectivity efforts and manage submissions to NGI as is done with other biometric submissions such as ten-prints. The transition process may also help establish a process for SIBs to establish connectivity to additional local agencies.

Overall, in an environment with constrained resources, vertical connectivity will be more feasible than horizontal connectivity across jurisdictions, because agencies can use the existing network infrastructure used for ten-print searches. State-to-state connectivity is of paramount importance, because it will allow for cross-jurisdictional searches that will benefit both state and local law enforcement activities. Connectivity to NGI through LITS-compliant AFIS systems will facilitate state-to-state connectivity.

Once an agency has a LITS-compliant system that is connected to a network, the next step is to facilitate searches against its AFIS. To do so, an agency may make its database more available to searching by registering with CJIS to receive a unique identifier for the Name of Designated Repository (NDR), a field specified in LITS. Once a data sharing MOU is in place, the NDR will facilitate sharing by allowing a law enforcement agency with a LITS-compliant AFIS system to add additional search database destinations into its system, without additional coding.

## Path Forward for Network Connectivity

The FBI's CJIS should work to expand CJIS-WAN to incorporate state-to-state connectivity, because it is already used by other Federal agencies and can handle a large bandwidth. The FBI should also look at measures to encourage more agencies to register their AFIS systems with the NDR so that they may be searched by other jurisdictions in relevant investigations. As more local agencies become LITS-compliant, the FBI could review its policies that restrict local agencies' access to NGI.

For some local jurisdictions, the benefit of local-to-local network connectivity may exceed the current challenges because of the cross-jurisdictional nature of crime in those locations. NIST, NIJ, and CJIS should develop guidelines on how to connect inter- and intra-state and local jurisdictions.

## Proper Governance

Governance agreements create a regulatory framework where cross-jurisdictional latent print searches are routine, standardized, secure, and compliant with relevant privacy policies. Once executed, generally by way of two or more agencies entering into an MOU, agencies with interoperability agreements can securely exchange electronic data within the bounds of privacy laws. By clarifying limits of use and standardizing sharing procedures, information exchange can advance from ad hoc and informal means to a more profound and integrated collaboration that reflects the complex nature of ensuring public safety today. These governance documents are best used when they are built on the foundation of technical interoperability and network connectivity as discussed in the previous subsections.

The current hierarchical nature of AFIS connectivity means that AFIS systems are integrated more vertically than horizontally and that the potential for horizontal collaboration has yet to be realized. Most local agencies have no formal sharing or governance agreements, even in areas of regional collaboration among states, which is the case for many of the local agencies in the WIN network. An exception is the NOVARIS agreement, which permits regional sharing in the Washington, D.C. area.[47] In response to NIJ's survey, approximately half of state agencies and a third of local agencies with an AFIS reported having an MOU or service-level agreement in place permitting them to use another AFIS.[48] This has resulted in a disparity in search capabilities and efficiencies among and between law enforcement agencies that have AFIS capabilities and a lack of clarity of when it is legal and appropriate to perform multi-jurisdictional searches. AFIS vendors often facilitate the development of MOUs, which means that these relationships often occur only where the agencies involved share the same vendor.

State and local laws regulate the sharing of civil and criminal fingerprint records collected under those jurisdictions. Many state and local jurisdictions have policies that preclude the enrollment of ten-print records obtained in a justice setting into another AFIS if the fingerprints originated from a minor or an individual convicted of a misdemeanor. Likewise, an agency may limit the retention of fingerprints after a certain period of time. These retention and sharing laws may not be consistent with those of neighboring jurisdictions. Better clarity around existing laws could help local law enforcement agencies develop joint policies that protect civil liberties and respect jurisdictions' preferences for privacy while facilitating sharing among agencies with common criminal pools.

Nationally, many of the privacy concerns are governed by the Federal Privacy Act of 1974, which allows for collection of fingerprints for civil or criminal law enforcement, counterterrorism efforts, and public safety matters. In addition, NGI users must abide by CJIS's security policy regarding "protecting the sources, transmission, storage, and generation of Criminal Justice Information (CJI)."[49]

The FBI's Compact Council, established in 1998 by the National Crime Prevention and Privacy Compact Act, has written rules regarding the collection and use of fingerprints for noncriminal justice purposes (such as application for a job or license, an immigration or naturalization matter, security clearance, or adoption).[50] Further research should be undertaken to assess the extent to which state laws vary in terms of collection of fingerprint data and subsequent use.

## Recent Progress on Governance and Remaining Challenges

To facilitate more standardized governance agreements, the NIST/NIJ Latent Print AFIS Interoperability Working Group drafted guidelines for developing an MOU between and among agencies that want to become interoperable.[51] The guidelines include a template for latent print processing agreements between two or more agencies, an overview of how to conduct the process, and sample structure and language for the agreement. The guidelines also provide sample language to address security and privacy disclosures. If an agency adheres to these guidelines, it can produce comprehensive and clear Standard Operating Procedures (SOPs) for cross-jurisdictional latent fingerprint searches.

In addition to the MOU guidelines, the Working Group published *Glossary of AFIS Terms*.[52] The glossary provides examiners and AFIS users with a comprehensive list of terms associated with AFIS technology. Not only does this glossary allow examiners to communicate more easily with one another using standard terminology, it also allows others involved with AFIS systems and development, such as procurement officers, vendors, and attorneys, to communicate more clearly.

The FBI is expected to publish a Privacy Impact Assessment on the privacy implications of NGI, including those relevant to latent services. This document is expected to address the additional privacy issues associated with retaining civil fingerprints and searching those prints against incoming criminal prints.

## Path Forward for Proper Governance

To encourage cooperation between and among state and local agencies, NIST should host a workshop or webinar series on how to implement the MOU guidelines. CJIS could develop strategies to coordinate the development of interstate local AFIS governance agreements.

The Department of Justice's Office of Justice Programs could conduct an analysis of variability in governance restrictions as a hurdle for interoperability. Such an analysis could begin with the collection and assessment of vertical and horizontal sharing policies, state and local privacy laws, and other policies related to the retention and use of fingerprints. This analysis could also examine best practices among biometric and other interoperable systems in developing policies and system specifications that govern federated access control and authorization systems in bidirectional, human readable, and machine-readable forms.

## Performance Testing and Training

Underpinning any interoperable system is the need for assurance that examiners are adequately trained and software complies with newly adopted standards, provides accurate results, and functions properly. Building in performance testing and training not only helps to maintain the integrity of criminal investigations pursuant to searches, but also serves to advance the field of fingerprinting and biometrics. Additionally, automated solutions must be put in place to assist the process of combining and sorting candidate lists so that examiners will have the time to follow up on candidate matches as a result of searching other systems. Inherently, the more data searched, the greater the number of candidates returned. Therefore, interoperability will generate a significant amount of work for examiners on the back end to review the increased number of candidates and will require a mechanism to support efficient prioritization and sorting of the many candidates. Moreover, proper training is needed to ensure that examiners are searching appropriately within the bounds of current privacy laws.

## Training and Tools

Latent fingerprint examiner training is a critical component of any latent print operation and is also an essential component of AFIS interoperability. Training will become ever more essential as technological upgrades are integrated into AFIS operations. For example, as jurisdictions upgrade their systems to become LITS conformant, latent print examiners will need to receive comprehensive training to ensure they are proficient in the use of the standard EFS markup and in determining which search profile is appropriate to use. Training can also reduce the risk of print comparison error and increase the standardization of the identification process.

In addition to basic fingerprint processing, examiners must be trained in methods for encoding the EFS latent print features and those required for vendor-specific proprietary AFIS database requirements. Unfortunately, there is no universal training program in the field of latent fingerprints. Agencies are typically responsible for designing and implementing their own training programs, and therefore, examiner training varies widely. While most agencies have some certified staff adjudicating candidate lists, most agencies do not require that all examiners be certified.[53] While some training is provided by the vendor, most training is conducted on the job through a mentoring program. The variability in the quality of mentoring is reflected in the variable quality of the submissions. For instance, among responding state agencies in the NIJ Survey, one-quarter of states had submissions rejected by the FBI because the submission did not meet the minimum quality standard.[54]

## Recent Progress on Training and Remaining Challenges

To respond to the demands for standardized training, NIST developed guidelines for EFS markup to provide "instructions for latent print examiners in marking friction ridge features to maximize consistency among examiners."[55] NIST, in cooperation with the FBI Biometric Center of Excellence, has also created an EFS online training tool as an interactive guide to latent fingerprint markup.[56] The tool provides an overview on the types of search profiles designated by LITS and allows examiners to practice annotating minutiae.

Further development of training should build on SOPs and proper management and administrative reporting. SOPs complement training programs by improving examiner performance, staff resource allocation, and examiner expectations. But many latent fingerprint operations do not have documented SOPs around searching other AFIS databases in their agencies.

SOPs should establish a framework for allocating examiner and AFIS resources and should include a policy for determining which searches to make depending on the available resources. The procedures should also outline document requirements for any latent fingerprint identification and should offer policies to reduce the risk of a print comparison error, the degree of review required, and the review processes required, based upon levels of comparison difficulty. Finally, the procedures should support the collection and analysis of operations data sufficient for system utilization and upgrade planning (e.g., number of searches made, cases solved, search characteristics, and success factors).

Training materials and programs should include methods for communicating to an examiner how to process latent fingerprints in accordance with established and validated SOPs.[57] Communications training will help demonstrate that the latent print comparison is accurate and admissible in court.

## Path Forward for Proper Training and Tools

The Federal Government should produce resources to help examiners efficiently and competently perform interoperable searches. To facilitate dissemination of best practices for LITS-compliant systems, NIST should develop guidelines for writing SOPs that incorporate reporting and best practices into the latent print identification process for searches conducted against other agencies' databases. NIJ or the Bureau of Justice Assistance (BJA) should introduce grant programs to support examiner training activities related to the implementation and use of LITS/EFS, to mitigate the costs of examiner retraining. In addition to these supporting efforts, NIST should develop guidelines on how examiners might communicate accurate results from a multi-jurisdictional search in testimony and reports.

Just as the FBI has provided the ULW freeware to process NGI search requests, NIST and the FBI should consider developing and making available an automated aggregation tool for candidate lists that are returned from multiple searches in one or more systems when simultaneous searches are appropriate. The tool should be able to prioritize the results and identify the most likely candidate from the combined searches.

## Testing

Testing AFIS systems helps ensure that they deliver accurate results while conforming to newly implemented transaction standards. Independent assessment of software performance would help ensure that users understand complex issues that underpin the technical aspects of interoperability and latent print identification. Testing could also be used to optimize examiner workflow, which would help address the increasing examiner demands caused by longer candidate lists as a result of enabling the searching of more AFIS systems. Workflow modifications should also account for the differences in algorithmic weighting that occur.

Testing of software has been incorporated into many of the vendors' business models, but not into law enforcement agency workflow. Independent verification of AFIS identification algorithms could help to ensure that vendors' claims about accuracy and reliability are not exaggerated to capture market share.

## Recent Progress on Testing and Remaining Challenges

Performance testing measures the accuracy of latent AFIS searches. Such testing will help ensure that reliability and accuracy objectives are not compromised when implementing interoperable standards. Agencies should require vendors to pass performance tests before accepting a delivered system. Performance testing has yet to be standardized, but the Latent Print AFIS Interoperability Working Group has identified a few elements that should be included in any performance test. NIST has employed testing to assess the accuracy of a print encoded on another AFIS system to a natively encoded print as a means of evaluating interoperability. NIST conducted two performance evaluations in 2011 and 2012 to assess the accuracy of "latent fingerprint searches using features marked by experienced human latent fingerprint examiners, in addition to automatic feature extraction and matching (AFEM)."[58] The evaluations also assessed which combinations of features were the most accurate search components, finding that "additional features resulted in accuracy improvement" in many cases.[59]

Federal testing efforts would be strengthened by independent third-party assessment of system architecture and performance. Independent assessment of performance and improvements in interoperability requires representative research data sets. These data sets would be especially beneficial if they included new forms of exemplar prints that are being collected by agencies. These emerging reference samples include prints and "slaps" from novel comparison sources such as mobile devices and 3D fingerprint renderings obtained from contactless fingerprint technologies.

## Path Forward for Testing

To ensure that interoperability standards do not compromise search accuracy, NIST could consider developing mechanisms to assess the accuracy of natively or remotely encoded prints and include language in the RFP guidance to ensure accuracy.

To encourage independent assessment and validation of system capabilities, NIST could develop and release a new research data set that includes new and novel comparison sources.

## Education and Outreach

The decentralized nature of American law enforcement introduces several challenges with dissemination of pertinent information relating to each of the core elements of interoperability. When standards, resources, policy updates, and training opportunities are adopted, developed, introduced, or announced, not all agencies and relevant stakeholders learn of these developments in a timely manner or at all. In particular, some state and local agencies are not aware of how the newly adopted LITS specification should be implemented, the availability of resources about the standard, or the changes to CJIS search and data retention policies. Education and outreach are important overarching components for bringing about interoperability.

Following the adoption of the LITS specification, the Latent Print AFIS Interoperability Working Group worked to develop guidelines for developing MOUs between agencies looking to enter into a latent print sharing agreement and for appropriate RFP language to ensure cost-effective LITScompliance in a new or upgraded AFIS system. Some state and local agencies have not been informed of these resources, jeopardizing the incorporation of LITS into the next generation of AFIS systems across the country.

When IAFIS was initially introduced, CJIS harbored concerns about exceeding capacity and placed a number of restrictions on what searches could be performed to manage the system. Augmented capacity and the introduction of NGI has enabled CJIS to lift these restrictions. Some state and local agencies maintain the perception that these restrictions are still in place today. This has led to confusion among state and local agencies as to when and under what conditions searches can be performed against NGI. For example, state or local police officers may be uncertain whether they are permitted to search NGI for a suspected leader of a drug syndicate charged with Federal crimes who is being held in a local jail. Outreach to local agencies to clarify CJIS's policies should be made a priority to ensure that law enforcement agencies can search NGI when their investigations call for it.[60]

CJIS staff members work closely with individual agencies to evaluate LITS and EBTS implementation in new AFIS databases. An expansion of these efforts could help translate technical specifications into practice by resource-constrained law enforcement agencies seeking to update their systems.

## SUMMARY OF RECOMMENDATIONS

The development of standards for latent print encoding and searching (EFS and LITS) is a major step forward in AFIS interoperability. Now the Federal Government must ensure that these standards are followed by helping state and local agencies build connections across jurisdictions, developing an overall connectivity strategy, fostering better governance, and supporting system-wide quality assurance through testing and training programs. Progress in each of these areas should be evaluated on an ongoing basis.

### Interagency Coordination

The Subcommittee on Forensic Science recommends that an interagency group be established to promote AFIS interoperability standards among Federal, state, and local authorities. This group would also be charged with coordinating and facilitating the development and adoption of standards related to AFIS use. Coordination of these activities by an interagency group, with cooperation of Federal, state, local, and tribal partners, as well as advice from industry, will help to ensure that users are provided with an integrated suite of standards to support law enforcement and counterterrorism needs. The suite of standards necessary to achieve interoperability and improve efficiency includes standards for evaluating equipment, training and certifying users, accreditation, and development of SOPs.

## Technical Compatibility

The Subcommittee recommends:

- All Federal AFIS should be fully compliant with LITS and EBTS without the use of an additional workstation within 3 years.
- The Federal Government should encourage adoption of LITS-compliant AFIS systems by state and local agencies:
  - NIJ or BJA should make funds available through existing grant programs to support state and local agency procurement of LITS-compliant AFIS.
  - All Federal funds made available to state and local agencies for AFIS procurement or upgrades should specify that the AFIS databases must be LITS-compliant.
- NIST should pursue activities to support the implementation of LITS-compliant AFIS systems through:
  - Periodic review of recently adopted standards;
  - Backward compatibility of the standards following any updates; and
  - Compliance through the development of conformance testing standards and funding conformance testing programs.

## Network Connectivity

The Subcommittee recommends:

- The FBI's CJIS should work to improve state-to-state connectivity by expanding the CJIS-WAN and encouraging more agencies to participate in development of a truly interoperable system.
- The FBI's CJIS should support local-to-state connectivity and local-to-local connectivity through technical assistance and develop guidelines with NIST and NIJ on how to interconnect local jurisdictions.

## Governance

The Subcommittee recommends:

- The FBI's CJIS should develop strategies to coordinate the development of interstate local AFIS governance agreements.
- NIST should host a workshop or webinar series on how to structure interagency agreements that allow for latent print searches that are routine, standardized, secure, and in compliance with relevant privacy policies.
- The Office of Justice Programs should conduct an analysis of horizontal and vertical sharing policies as well as state and Federal privacy laws and policies that impact fingerprint sharing across jurisdictions and identify opportunities to coordinate the development of joint policies.

## Performance Testing and Training

The Subcommittee recommends:

- The Federal Government should support training activities of LITS-compliant systems through the following activities:
    - NIST should develop support materials, including SOPs for examiners.
    - NIJ and BJA should make funds available through current grant programs to support examiner training activities related to the implementation and use of standards-compliant systems.
- NIST should develop performance tests to assess the matching accuracy of natively versus remotely encoded prints and new research data sets to encourage independent testing.
- NIST and the FBI should develop and make available an automated tool to aggregate candidate lists from multiple searches in one or more systems that could lead to improvements in efficacy and accuracy.

## Education and Outreach

The Subcommittee recommends:

- The Federal Government should help resource-constrained law enforcement agencies seeking to update their systems by expanding efforts to evaluate LITS and EBTS implementation and translate technical specifications into practice.

# REFERENCES

Hicklin, Austin. *Standardizing a More Complete Set of Fingerprint Features.* Noblis, Inc. 2007. Prepared for the Committee to Define an Extended Fingerprint Feature Set. Accessed 9 July 2014. www.fingerprint.nist.gov/STANDARD/cdeffs/Docs/IAI CDEFFS 2007-07-24.pdf.

Indovina, M., R. A. Hicklin, and G. I. Kiebuzinski. "ELFT-EFS Evaluation of Latent Fingerprint Technologies: Extended Feature Sets, Evaluation #1." NISTIR 7775. March 2011.

Indovina, M., V. Dvornychenko, R. A. Hicklin, and G. I. Kiebuzinski. "ELFT-EFS Evaluation of Latent Fingerprint Technologies: Extended Feature Sets, Evaluation #2." NISTIR 7859. May 2012.

Landon, James J. *Privacy Impact Assessment Integrated Automated Fingerprint Identification System (IAFIS)/Next Generation Identification (NGI) Repository for Individuals of Special Concern (RISC).* Federal Bureau of Investigation, July 10, 2012. Accessed July 8, 2014. www.fbi.gov/foia/privacy-impact-assessments/iafis-ngi-risc.

Moses, Kenneth R., Peter Higgins, Michael McCabe, Salil Probhakar, and Scott Swann. "Chapter 6—Automated Fingerprint Identification Systems (AFIS)." In *The Fingerprint Sourcebook.* Washington, D.C.: Department of Justice, 2010: 6-1–33.

National Institute of Standards and Technology (NIST). *Latent Interoperability Transmission Specification.* NIST Special Publication 1152. January 2013. doi: 10.6028/NIST.SP.1152.

———. *American National Standard for Information Systems—Data Format for the Interchange of Fingerprint, Facial, & Scar Mark & Tattoo (SMT) Information.* NIST Special Publication 500-245, ANSI/NIST-ITL 1-2000. Washington D.C.: NIST, July 2000.

———. *Data Format for the Interchange of Fingerprint, Facial & Other Biometric Information.* NIST Special Publication 500-290, ANSI/NIST-ITL 1- 2011. November 2011.

———. *Data Format for the Interchange of Fingerprint, Facial & Other Biometric Information.* NIST Special Publication 500-290 Rev1, ANSI/NISTITL 1-2011 Update: 2013. December 2013.

———. *Markup Instructions for Extended Friction Ridge Features.* NIST Special Publication 1151. January 2013. doi: 10.6028/NIST.SP.1151.

———. *Writing Guidelines to Develop an MOU for Interoperable Automated Fingerprint Identification Systems.* NIST Special Publication 1156. May 2013. doi: 10.6028/NIST.SP.1156.

National Institute of Standards and Technology (NIST), Latent Print AFIS Interoperability Working Group. *Glossary of AFIS Terms.* February 17, 2012.

———. *Writing Guidelines for Requests for Proposals for Automated Fingerprint Identification Systems.* February 17, 2012. Noblis, Inc. *Latent Print Interoperability: State and Local Perspectives.* April 2, 2012.

Persinger, Mark, Lars Ericson, and Mark Greene. *Latent Fingerprint Interoperability Survey: A National Study of Automated Fingerprint Information Systems (AFIS) Maintained by Law Enforcement Agencies. Summary Reporting of Data Provided by Responding Agencies.* Washington, D.C.: Department of Justice, August 2014.

U.S. Department of Justice, Federal Bureau of Investigation (FBI). *Electronic Biometric Transmission Specification (EBTS) Technical and Operational Update (TOU) 10.0.2.* (June 2, 2014). Accessed August 21, 2014. www.fbibiospecs.org/docs/EBTS%20TOU%201002Final.pdf.

———. *Privacy Impact Assessment Integrated Automated Fingerprint Identification System (IAFIS)/Next Generation Identification (NGI) Repository for Individuals of Special Concern (RISC).* July 10, 2012. Accessed September 2, 2014. www.fbi.gov/foia/privacy-impact-assessments/iafis-ngi-risc.

U.S. Department of Justice, Federal Bureau of Investigation (FBI). "Uniform Crime Reports: Crime in the United States, 2012." Fall 2013. Accessed July 7, 2014. www.fbi.gov/about-us/cjis/ucr/crime-in-the-u.s/2012/crime-in-the-u.s.-2012/offenses-known-to-lawenforcement/clearancetopic.pdf.

———. *Criminal Justice Information Services (CJIS) Security Policy.* Criminal Justice Information Services, August 4, 2014. www.fbi.gov/aboutus/cjis/cjis-security-policy-resource-center/at download/file.

U.S. Department of Justice, Federal Bureau of Investigation (FBI), Compact Council. "Agency Privacy Requirements for Noncriminal Justice Applicants." Available at www.fbi.gov/about-us/cjis/cc/library/agency-privacy-requirements-for-noncriminal-justice-applicants.

## End Notes

[1] EFS refers to the definitions of fingerprint and palm print features incorporated in the ANSI/NIST-ITL standard starting in 2011. (U.S. Department of Commerce, National Institute of Standards and Technology (NIST), Latent Interoperability Transmission Specification (LITS), NIST Special Publication 1152, January 2013.) The FBI's Electronic Biometric Transmission Specification is compliant with LITS. (U.S. Department of Justice, Federal Bureau of Investigation, Electronic Biometric Transmission Specification (EBTS) Technical and Operational Update (TOU) 10.0.2, June 2, 2014.)

[2] National Institute of Standards and Technology (NIST), Latent Interoperability Transmission Specification, NIST Special Publication 1152, January 2013.

[3] U.S. Department of Justice, FBI, Electronic Biometric Transmission Specification (EBTS) Technical and Operational Update (TOU) 10.0.2.

[4] NIST, Markup Instructions for Extended Friction Ridge Features, NIST Special Publication 1151, January 2013. The training tool is available at www.nist.gov/ forensics/ EFSTrainingTool.

[5] M. Indovina, R. A. Hicklin, and G. I. Kiebuzinski, "ELFT-EFS Evaluation of Latent Fingerprint Technologies: Extended Feature Sets, Evaluation #1," NISTIR 7775, March 2011; M. Indovina, V. Dvornychenko, R. A. Hicklin, and G. I. Kiebuzinski, "ELFT-EFS Evaluation of Latent Fingerprint Technologies: Extended Feature Sets, Evaluation #2," NISTIR 7859, May 2012.

[6] U.S. Department of Justice, FBI, "Offenses Cleared," Uniform Crime Reports: Crime in the United States, 2012, Fall 2013.

[7] The Henry Classification System was in place through the 1990s, and may still be used today by agencies without AFIS databases.

[8] Early AFIS users continued to use the manual Henry Classification System. In the 1980s, AFIS databases were developed to meet the growing demands for identification and began incorporating functionality to search latent fingerprints against a complete database. This increased capacity became a major selling point to the fingerprint community.

[9] U.S. Department of Justice, FBI, Privacy Impact Assessment Integrated Automated Fingerprint Identification System (IAFIS)/Next Generation Identification (NGI) Repository for Individuals of Special Concern (RISC), July 10, 2012.

[10] Kenneth R. Moses, Peter Higgins, Michael McCabe, Salil Probhakar, and Scott Swann, "Chapter 6—Automated Fingerprint Identification Systems (AFIS)," in The Fingerprint Sourcebook (Washington, D.C.: U.S. Department of Justice, 2010): 6-1–33.

[11] When first introduced, EBTS was known as the Electronic Fingerprint Transmission Specification.

[12] NIST, Latent Interoperability Transmission Specification.

[13] Reasons beyond interoperability initiatives result in IAFIS not mirroring other ten-print databases. Some of these reasons involve policy or legal issues, such as lack of data sharing policies.

[14] Mark Persinger, Lars Ericson, and Mark Greene Latent Fingerprint Interoperability Survey: A National Study of Automated Fingerprint Information Systems (AFIS) Maintained by Law Enforcement Agencies. Summary Reporting of Data Provided by Responding Agencies (hereafter referred to as LFIOS) (Washington D.C.: Department of Justice, August 2014), Local Addendum Question 28 and State Addendum Question 8.

[15] LFIOS, State Addendum Question 8.

[16] LFIOS, Question 9.

[17] Among responding local law enforcement agencies, 43% do not have a system compatible with their State AFIS, and 45% have an AFIS made by the same vendor as the State. LFIOS, Local Addendum Questions 4 and 5. According to Question 1 of the State Addendum of LFIOS, the local law enforcement agencies of 14 states do not maintain independent AFIS databases.

[18] LFIOS, Local Addendum Question 16.

[19] This figure was consistent across both state and local responding agencies (LFIOS, Questions 162 and 63).

[20] While this would suggest that 31 agencies routinely searched NGI at the time of data collection, the FBI was still using IAFIS (LFIOS, State Addendum Questions 30 and 31).

[21] LFIOS, Local Addendum Question 39.

[22] LFIOS, Question 169 and Question 170.

[23] Kansas Bureau of Investigation, New Hampshire, Vermont, North and South Dakota, Minnesota, Connecticut, Rhode Island, Las Vegas, and other jurisdictions have also employed regional interoperability approaches.

[24] The Office of Biometric Management was formerly known as United States Visitor and Immigrant Status Indicator Technology.

[25] The USA PATRIOT Act of 2001 (P.L. 107-56) required the Attorney General and the FBI to make available to the Department of State and the INS (now the U.S. Citizenship and Immigration Services) records for determining whether or not a visa applicant or applicant for admissions has a criminal history.

[26] The Homeland Security Act of 2002 (P.L. 107-296) required DHS to oversee and coordinate DHS programs for and relationships with state and local governments.

[27] The Enhanced Border Security and Visa Entry Reform Act of 2002, (P.L. 107- 173) required an interoperable law enforcement and intelligence data system and the ability to update IDENT with known and suspected terrorist information.

[28] The Intelligence Reform and Terrorism Prevention Act of 2004, (P.L. 108-458) required creation of an information sharing environment to share terrorism information across the homeland security, law enforcement, and intelligence communities at all levels of government and the private sector. It also accelerated development of an integrated electronic biometric entry-exit screening system (i.e., the Entry/Exit system).

[29] LFIOS, State Addendum Questions 51 and 54.

[30] The ANSI/NIST-ITL standard underwent minor revisions in 2013. Localities including LITS in their contracts are Orange County, California, and the WIN.

[31] For states and local agencies that responded that they had upgraded their systems (LFIOS, Question 25), this average included the year they had upgraded (Question 26). For those that had not upgraded their AFIS, the average included the year the system became operational (LFIOS, Question 11).

[32] LFIOS, Question 24 and Question 35.

[33] When authorized, examiners may search latent prints against civil records.

[34] James J. Landon, Privacy Impact Assessment Integrated Automated Fingerprint Identification System (IAFIS)/Next Generation Identification (NGI) Repository for Individuals of Special Concern (RISC) (Federal Bureau of Investigation, July 10, 2012).

[35] An environment of distinct systems may encourage better search algorithms through vendor competition. Models, however, that directly encourage vendors to compete for a single procurement, such as those employed by the Indian Government, may be especially effective at spurring advancing AFIS accuracy.

[36] At the ANSI/NIST ITL 1-2000 Standard Workshop I in April 2005, the Scientific Working Group on Friction Ridge Analysis, Study, and Technology was tasked to identify, define, and provide guidance on additional fingerprint features beyond the traditional ending ridges and bifurcations defined in the ANSI/NIST ITL 1-2000 standard. (NIST, Data Format for the Interchange of Fingerprint, Facial, & SMT Information.)

[37] NGI interoperability is ensured because LITS is compliant with EBTS, which includes specifications for many biometric modalities. EBTS specifications can be found at www.fbibiospecs.org/ebts.html.

[38] NIST also developed resources instructing examiners how to properly annotate features using EFS, including the Extended Feature Set Training Tool found at www.nist.gov/forensics/EFSTrainingTool/TrainingTool.html. (Indovina, Hicklin, and Kiebuzinski, "ELFT-EFS Evaluation #1"; Indovina, Dvornychenko, Hicklin, and Kiebuzinski, "ELFT-EFS Evaluation #2.")

[39] For the full specification see noblis.org/media/d58f0f47-37b9-4ea2-81c5-ab5a6972f62f/docs/LITSv1-02012-02-15pdf and nvlpubs.nist.gov/nistpubs/SpecialPublications/NIST.SP.1152.pdf.

[40] Image-only searches can be supplemented with a minimal markup profile for the purposes of defining a region of interest, orientation, finger/palm print position, pattern class, cores, and deltas.

[41] NIST, Latent Print AFIS Interoperability Working Group, Writing Guidelines for Requests for Proposals for Automated Fingerprint Identification Systems, February, 17 2012.

[42] LFIOS, Question 25.

[43] LFIOS, Question 35.

[44] LFIOS, State Addendum Question 43. Because most states receive results from the ULW, it may be worth considering whether or not the ULW should be expanded to include "plug and play" proprietary algorithms for search results and to enable latent examiners to have a consistent graphical user interface across agencies.

[45] LFIOS, State Addendum Question 38.

[46] LFIOS, Questions 242 and 243.

[47] Even among NOVARIS's partners, examiners must re-encode prints to search their respective state AFIS and therefore have limited access to the NGI.

[48] LFIOS, Question 241.

[49] U.S. Department of Justice, FBI, Criminal Justice Information Services (CJIS) Security Policy, (Criminal Justice Information Services, August 4, 2014).

[50] U.S. Department of Justice, FBI, Compact Council, "Agency Privacy Requirements for Noncriminal Justice Applicants."

[51] NIST, Writing Guidelines to Develop an MOU for Interoperable Automated Fingerprint Identification Systems, NIST Special Publication 1156, May 2013.

[52] NIST, Latent Print AFIS Interoperability Working Group, Glossary of AFIS Terms, February 17, 2012.

[53] LFIOS, Question 270 and 271.

[54] LFIOS, State Addendum Question 40.

[55] NIST, "Markup Instructions for Extended Friction Ridge Features."

[56] The NIST training tool is available at www.nist.gov/forensics/EFSTrainingTool.

[57] The Scientific Working Group on Friction Ridge Analysis, Study and Technology previously developed model Standard Operating Procedures regarding latent print examinations that can be found at www.swgfast.org/Documents.html.

[58] Indovina, Hicklin, and Kiebuzinski, "ELFT-EFS Evaluation #1."

[59] Indovina, Dvornychenko, Hicklin, and Kiebuzinski, "ELFT-EFS Evaluation #2."

[60] FBI, CJIS Security Policy.

# INDEX

## #



## A



## B



## C

## D



## E

## J



## L



## M

## S



## T

## U



## V



## W



## X



## Y